T0308441

Threats

Threats

Intimidation and Its Discontents

David P. Barash

OXFORD
UNIVERSITY PRESS

OXFORD
UNIVERSITY PRESS

Oxford University Press is a department of the University of Oxford. It furthers
the University's objective of excellence in research, scholarship, and education
by publishing worldwide. Oxford is a registered trade mark of Oxford University
Press in the UK and certain other countries.

Published in the United States of America by Oxford University Press
198 Madison Avenue, New York, NY 10016, United States of America.

Library of Congress Cataloging-in-Publication Data
Names: Barash, David P., author.
Title: Threats : Intimidation and Its Discontents / by David P. Barash.
Description: New York, NY : Oxford University Press, [2020] |
Includes bibliographical references and index.
Identifiers: LCCN 2020004434 (print) | LCCN 2020004435 (ebook) |
ISBN 9780190055295 (hardback) | ISBN 9780190055318 (epub)
Subjects: LCSH: Threat (Psychology) | Threat (Psychology)—Social aspects. |
Danger—Psychological aspects.
Classification: LCC BF575.T45 B37 2020 (print) | LCC BF575.T45 (ebook) |
DDC 303.3/6—dc23
LC record available at https://lccn.loc.gov/2020004434
LC ebook record available at https://lccn.loc.gov/2020004435

1 3 5 7 9 8 6 4 2

Printed by Sheridan Books, Inc., United States of America

In a dark time, the eye begins to see,
I meet my shadow in the deepening shade.
—Theodore Roethke

Not everything that is faced can be changed, but nothing can be changed unless it is faced.
—James Baldwin

Only two things are infinite, the universe and human stupidity, and I'm not sure about the former."
—Albert Einstein

Contents

Acknowledgments

This book could not have happened without the wisdom and energetic enthusiasm of my agent, Howard Morhaim, and the benevolent efforts of my extraordinarily capable editor at Oxford, Jeremy Lewis. Bronwyn Geyer and Preetham Raj ably shepherded it through a wonderfully smooth production process. I am also very grateful to the following scholars who read various parts of the text and made (non-threatening!) comments and suggestions—most of which I followed: Joe Cirincioni, Richard Dawkins, Daniel Ellsberg, Rob Green, Susan Jacoby, Rick Shenkman, and Carol Tavris. I've always wanted to write that the people being acknowledged are responsible for anything wrong in the final book and that I, as author, deserve full credit for whatever is right . . . but of course, the opposite is the case! Finally, it is a particular pleasure to acknowledge my wife, Judith Eve Lipton, and to point out that, as ever, it is literally impossible to identify which of "my" ideas are actually "ours." And to hope that at least some of them will become yours.

A Nonthreatening Start

THREATS ARE UBIQUITOUS. An old Noel Coward song proclaims that "Birds do it, bees do it/Even educated fleas do it. . . . Let's fall in love." For our purposes, the "it" is different: giving and receiving threats, trying to make it clear that if someone does something unwanted, the consequences—typically some form of retaliation—will be that the potential perpetrator won't perpetrate in the first place. Threats are so encompassing they often go unnoticed, just as the hypothetical intelligent fish who, when asked to describe her world, is unlikely to reply, "It's very wet down here." Threats are the ocean in which we swim. The goal of this book is to make the water visible.

Sometimes threats are blindingly obvious: a raised fist, a pointed gun, or an explicit written warning—KEEP OFF THE GRASS! FINES DOUBLED DURING SCHOOL HOURS! Other threats are ostensibly subtle but nonetheless clear, as in *The Godfather*'s "Make him an offer he can't refuse," or mostly unstated but resonant, as with nuclear weapons poised for retaliation. They are also exquisitely sensitive to context, as when "The Virginian," in Owen Wister's 1902 novel of that name, responds to being labeled a son-of-a-bitch by intoning, "When you call me that, smile." Or the difference between hearing, "Lovely child you have there" and "It would be a shame if something bad happened to her," said by a mob boss or by a pediatrician recommending a measles vaccine.

Although we are especially aware of explicit threats, such as "If you do X, then I'll do Y," others are often hidden, leaving most of the interpersonal ones implicit and therefore psychologically fraught. Warnings, by contrast, can be impersonal and not necessarily conveyed by an individual: "SMOKING IS HAZARDOUS TO YOUR HEALTH," or CAUTION: DANGEROUS UNDERTOW. But sometimes, threats and warnings are indistinguishable: "Violators will be prosecuted." We worry, of course, about threats to human life and safety that are beyond our control—earthquakes, pandemics, floods, tsunamis, volcanic eruptions, and other large-scale disasters—but nearly everyone is more agitated by those emanating from fellow humans. More people in the United States are killed annually by auto collisions with deer or by falling refrigerators and vending machines than by terrorists, but it is the "terrorist threat" that evokes particular public anxiety as well as military action. At least within the human psyche, people are always high in the threat hierarchy, although—we'll see—God is no piker in this respect.

In her book, *The Big Ones*, seismologist Lucy Jones puckishly points out that "shift happens."[1] An appropriate warning from a geologist, but human agency has its own unique qualities. A salient point in *Threats: Intimidation and its Discontents* is that although most "shifts" aren't affected by whether people perceive them as threatening, many human responses to person-generated threats actually make things worse, when a perception of *being* threatened induces a response or counterthreat, leaving everyone worse off than before—for example, gun ownership in response to threats of violence, "antivaxxer" parents responding to what they see as a threat posed by immunizations and who thereby increase the risks to their own children along with everyone else's, nuclear weapons as a supposed way of fending off threats to national security. And more.

Undergirding these problems are how human threats often connect with irrationality. Paradoxically, threats —and responses to them—in the world of plants and animals are nearly always powered by a deep logic, not so much generated consciously by the organisms themselves as by natural selection, a relentless mathematical process whereby success is reproductively rewarded and failure punished. Although human beings like to congratulate themselves for their cognition and rationality, we all know that emotion, drives, and irrationality often carry the day. Nor is this always bad; however, when it comes to threats, both our broadcasting and our responding have a sorry history, often because of failure to cognitively over-ride our less rational inclinations. So, section 1 considers the surprisingly "rational" use of threats by non-human organisms, after which section 2 looks at frequent failures of rationality when it comes to how *Homo sapiens* issues and responds to our own perceived threats. And then, section 3 deals with an immense threat that we have created for ourselves, one that has been peculiarly encased in layers of faux rationality but which is fundamentally irrational in the extreme.

In *Civilization and Its Discontents*, Freud argued that human beings require civilization and yet, civilization demands that we inhibit important parts of ourselves, namely our penchant for violence and unmanaged sex. Intimidation, on the other hand, is not something that we require, but that warrants a great deal of discontent. And more attention than it has received so far.

This book was finalized when the coronavirus crisis was beginning to transfix the United States and the world. Given that the unfolding covid-19 catastrophe is essentially a natural event, analogous to a massive earthquake or tsunami, I decided to refrain from discussing it or its implications, focusing instead on various manifestations of intimidation—its reality and its discontents.

We'll start by going through the biology of threats as conveyed and received by animals, followed by a variety of socially mediated threats intended to prevent crime (notably, the use of capital punishment), along with such odious, consequential threats as hellfire and brimstone, personal death, gun violence, and the rise of right-wing national populism, culminating in a critical look at nuclear deterrence, the premier manifestation of threat at the international level. Of particular note is how responses to threats often become counterproductive: hardly at all in the case of animals, occasionally (but enough to be troublesome) in the personal and social realm, and then overwhelmingly when it comes to deterrence.

THE CLAUSEWITZ COP-OUT

Carl von Clausewitz warned that by seeking to defend everything, we risk defending nothing. Similarly, if this book were to seek to cover all threats, it would fail at any of them, if only because they are so abundant: perceived threats *of* immigration; genuine threats *to* immigrants; geopolitical and economic threats from China; threats of political meddling by Russia; existential threats to human and animal health from environmental destruction, pollution, and invasive species; the devastating opioid epidemic; pandemics such as covid-19, dangers of runaway biotechnology and artificial intelligence; challenges to the biological and physical integrity of Earth itself. Alas, the list is long and, the preceding, woefully incomplete. In what is to come, some of these threats are gestured at, while others are ignored completely, not because they're unimportant, but because Clausewitz was right. Rather than an encyclopedic account of all threats, our look at intimidation and its discontents is therefore more than a little biased. So many disparate things can be folded within the rubric of threats that any selection is necessarily idiosyncratic.

Many of the threats that here go unexplored are important indeed, but in most cases they have been discussed at length and with greater authority and depth by others. Threats of financial and physical harm have, for example, been a mainstay of domestic abusers, whereas anticipated physical harm is a major component of the psychological torture endured by victims of schoolyard bullying: "I'm gonna beat you up after school . . . . ! For a different form of bullying (whose effects are also likely to be hurtful), consider the pronouncement by Nikki Haley, ex-US ambassador to the United Nations, that if member nations don't vote as the United States wants, their aid will be cut off; or the similar threat from President Trump that the Palestinians won't receive humanitarian assistance if they don't vote in favor of whatever his administration demands.

A much-discussed contribution to the 2016 election victory of Donald Trump was the degree to which white working-class men in particular felt threatened by their loss of social control and economic security, exacerbated by the influx of immigrants who, in their judgment, threaten control of "their" country, not to mention their jobs. Ironically, a president elected—to a substantial degree—because many voters felt threatened became unique in American history for his divisive use of threats in political discourse. He became a unique threat in his own right: to decency, democracy, the environment, independence of the press, and of the federal judiciary. He trampled on basic standards of civility, norms of ethics, honesty, and personal integrity; the rule of law; and the importance and credibility of science, and of America's international standing, including diminished reliability vis-à-vis friends and allies—combined with a perverse embrace of dictators worldwide, even seeking to despoil the legitimacy of facts themselves. To describe Donald Trump as a solution to the threats faced by the United States and the world is like describing gasoline as a solution to a fire. Nonetheless, this book invokes author privilege and focuses considerably less on Mr. Trump than he deserves, simply because that bleak territory has been well covered by others.

Short shrift—actually, no "shrift" (whatever that is)—is also given to the immense ongoing threat of global climate change, mostly because there have been so many superb treatments of this key topic[2] that perhaps now it is time to *do* more than to write. Ditto for economic competition from China and political meddling by Russia, each of which, once again, has generated its share of anxiety and corresponding attention. Nor do we consider pandemics, the statistically unlikely but potentially devastating effects of eruption by the roughly twenty currently dormant supervolcanoes scattered throughout our planet, an asteroid impact, the danger that runaway artificial intelligence could succeed in multiplying itself to our detriment, or the risk that we might someday be contacted—and then robbed, enslaved, or obliterated—by an advanced alien civilization.[3] This last one might seem especially far-fetched, but no less a mastermind than Stephen Hawking advised that we would do well to keep our heads down and refrain from searching for extraterrestrial intelligence, a warning developed in terrifying detail in the sobering Chinese science fiction trilogy widely known as *The Three-Body Problem*.

There has recently been substantial growth of concern with "existential risk," by which is meant those things that could threaten the existence of life on Earth and of human life in particular. For example, Cambridge University hosts the Centre for the Study of Existential Risk (CSER), and the Boston area with its powerful university community is home to the Future of Life Institute (FLI). Neither of these organizations qualify as Chicken Little, the-sky-is-falling fringe groups: the CSER's faculty includes cosmologist Martin Rees, and the FLI was founded by physicist Max Tegmark. It is sobering to consider that most of the major current threats to humanity and to planet Earth—nuclear war, climate change, out-of-control artificial intelligence, resource depletion, global ecosystem collapse, supervolcanoes, asteroid impacts—were not even on the conceptual horizon of professional worriers and futurists just a century or so ago. Assuming that, by wisdom or luck, we are able to duck these dangers, what new threats, currently unimagined, will our grandchildren and their descendants face?

In *Threats: Intimidation and its Discontents*, although much is excluded, much remains. For now, our look at the use and abuse of threats takes place in three sections.

WHAT'S TO COME

In Section 1, "The Natural World," we look at plants with thorns and poisons; animals growling, roaring, bearing teeth, showing and exaggerating their weapons (or pretending to have weapons), revealing and sometimes exaggerating their ferocity, puffing themselves up, and generally seeking to intimidate their rivals or potential predators. We also examine the role of honesty versus deception, the evolution of warning coloration (whereby bright-colored poison arrow frogs, for example, inform would-be predators that eating them would be a bad idea) as well as mimicry—when animals that are not themselves especially dangerous (e.g., the viceroy butterfly) resemble others (e.g., monarchs) that are harmful to their predators and gain protection via the "empty threat" the former conveys. This, in turn, speaks to the intriguing question of whether a given threat is real or fake, honest or dishonest, and what difference—if any—this makes. In addition, we eavesdrop on the importance of eavesdropping.

In Section 2, "Individuals and Society," we take a look at crime and punishment—notably, how criminal statutes seek to prevent crime by threatening criminals with punishment intended to provide an effective deterrent. There is a long and fascinating history of such efforts, with very little success. This leads us to consider the death penalty in particular and whether it has been effective in preventing capital crimes (spoiler alert: it hasn't). Included as well is a review of how people turn to religion when under threat, along with how religions have often threatened their adherents with after-death retribution for sin, which has long backgrounded much human anxiety and, in some cases, compliance. We also encounter—or at least, read about—the menace of death plus threats involved in the American gun culture, along with race-based and economic anxieties driving the rise of right-wing national populism.

In Section 3, "International Affairs," we examine how the use of military threats—that is, deterrence—has operated among nations, starting with a brief history of conventional deterrence aimed at preventing wars by threatening either that an initial attack will fail to gain its objective ("deterrence by denial") or that an attacker will be sufficiently punished so as to regret the initial action and, thus, prevent such an attack in the first place ("deterrence by punishment"). This leads to the self-serving phenomenon of bureaucratic "threat inflation" and then to the most consequential use and abuse of threats: nuclear deterrence, how it is supposed to work, whether it has in fact worked, and a critical analysis of its strengths and especially its weaknesses. The threat inherent in nuclear weapons is our deepest and darkest shadow, although Theodore Roethke did not seem to have it in mind when composing the words that appear in this book's epigraph. The nuclear shadow, however, is dark indeed; perhaps our eyes will eventually begin to see, and also, as James Baldwin pointed out, to change it once we honestly face it.

Last, this book ends with two brief appendices, the first a dire warning by a senior Air Force general and former head of US strategic nuclear forces, and the second a personal account of acceptable deterrence on the author's farm provided by a large, territorial, but altogether non-nuclear if unconventional dog.

NOTES

1. Jones, L. 2018. *The Big Ones: How Natural Disasters Have Shaped Us (and What We Can Do about Them)*. New York, N.Y.: Doubleday.
2. Wallace-Wells, D. 2019. *The Uninhabitable Earth*. New York, N.Y.: Tim Duggan Books; McKibben, B. 2019. *Falter: Has the Human Game Begun to Play Itself Out?* New York, N.Y.: Henry Holt; Rich, N. 2019. *Losing Earth: A Recent History*. New York, N.Y.: MCD (Farrar, Straus & Giroux).
3. For a plausible and therefore worrisome account of these and others, see Walsh, B. 2019. *End Times: A Brief Guide to the End of the World*. New York, N.Y.: Hatchette.

Section 1

The Natural World

THREATS, COUNTERTHREATS, WARNINGS, feints, and deception are found through-out the natural world, in the daily lives of animals and even plants. Roses and black-berries have thorns—saying, in effect, "Don't touch me . . . or else!" Confront a spider, perhaps one that has accidentally strayed indoors and is thus on human turf rather than its own, and the tiny creature will likely rear back on its hindmost legs and assume a threatening posture, one that is ridiculous given that it can easily be squashed with a shoe. Yet, everyone understands the gesture, although to locate the most recent com-mon ancestor shared by a spider and a human being one must go back more than half a billion years.

Threats have an ancient pedigree in the human imagination too. Among the most iconic and oft-repeated tales are explicit prohibitions: in the ancient Hebrew Bible, Yahweh warns Adam and Eve not to eat of the fruit of the Tree of Knowledge of Good and Evil, lest you "surely die." (They do anyhow.) Prometheus gives fire to human beings, despite Zeus's explicit order to the contrary. Also from Greek mythology, Pandora gets a box (originally, a jar) that she is forbidden to open. Bluebeard's young wife is similarly warned not to open a particular basement door. The list goes on—and the threats inevitably fail. A case can be made that, in the human context at least, folk-tales involving threats exist specifically to emphasize either the uselessness of making them or the foolishness (and frequency) of ignoring them.

DON'T TREAD ON THEM

There is no doubt that, in some cases, threats work—notably among animals. Lions have huge canines, not only for killing their prey, but also for threatening other lions. By the same token, lions are notably silent when stalking zebras, reserv-ing their impressive roar—along with showing off their magnificent dentition—to discourage other lions from invading their pride. When threatening another ani-mal, standard procedure is for "threateners" to make themselves seem larger, more imposing, more dangerous than they really are in an effort to deter an opponent from taking their food, nest site, mate, or, quite simply, from attacking. An analysis of the biology of animal threat signals pointed out that conflicts are "likely to be won by the individual that is larger, stronger, healthier, more experienced and/or more motivated. Threat signals are designed to transmit information about these questions."[1] Not surprisingly, the information thereby transmitted nearly always makes the threatener seem larger, stronger, healthier, more experienced, and more motivated than it really is—all in the service of deterring trouble, which is far less costly than actual combat. On the other hand, for the threat to work, threateners must signal that they have weapons and are willing and able to use them. (Even if neither is true.) And, as we shall see, there is a tendency among animals as well as human beings to exaggerate both their danger to others and the probability that any transgressions will be severely punished.

One of the most potent messages delivered by monkeys and apes is an unflinching "stare threat," typically perceived as intimidating not only by other conspecifics, but by human observers as well. Yet another threat display involves an erect penis, a widespread maneuver among many nonhuman primates, but in our own species, not so much—at least not, fortunately, in modern Western civilization. Why this particular threat signal ever existed is something of a mystery, given that the male organ, per se, is not an effective weapon and it appears that no one has ever been injuriously beaten with one. On the other hand, maleness in human beings—as in many other animals in general and mammals in particular—has long been associated with violence, so it is not unexpected that an unambiguous symbol of male sexuality would also indicate such a capability.

It is noteworthy that many people consider animals a threat, probably because *Homo sapiens* carries a deep-seated memory of when we were liable to be prey no less than predators. Thus, nonhikers in particular express anxiety about bears and cougars, although getting lost and suffering either heatstroke or hypothermia are far greater threats to survival in the outdoors. According to the Centers for Disease Control and Prevention, American deaths from bears, alligators, and sharks average about one per year for each species, with the mortality from cougar attacks being fewer than one per seven years. Venomous snakes and lizards combined kill roughly six per year; spiders kill seven; cattle—notably, bulls—kill twenty (nearly all farm workers); dogs, twenty-eight; and bees, wasps, and hornets, fifty-eight combined (because of allergic shock reactions). By contrast, motor vehicles kill thirty-three thousand people annually, yet nearly everyone shares with Dorothy, from *The Wizard of Oz*, more fear of lions, tigers, and bears—"oh my!"—than of Chevrolets.

Nature isn't, as Tennyson wrote, mostly "red in tooth and claw." Nevertheless, living things are typically equipped with a range of intimidating options: not just teeth and claws, but also horns, antlers, poison fangs, bony shields, scary hisses, and formidable roars, often wrapped up in a package designed to make themselves seem as ferocious and intimidating as possible. All the better to frighten you with, my dear.

At the same time, those issuing such threats must be prepared that they may fall short of their intended effect—in other words, "if deterrence fails" (a frequent and terrifying phrase within the community of nuclear strategists), something that occurs in nature, albeit less often than hyperventilating animal films and YouTube videos would suggest. It is relatively rare—although more common than initially claimed by Nobel Prize-winning ethologist Konrad Lorenz, in particular—for animals to kill or even seriously injure each other should their threats fail. Even if push comes to shove, antler to antler, beak to beak, tooth to tooth, or claw to claw, and if as a result there are winners and losers, the former generally get to enjoy the fruits of their victory while the latter, albeit defeated and disappointed, slinks, slithers, flies, climbs, or runs away—to fight, or at least threaten, another day. Often, the result is a standoff, with the would-be aggressor held at bay, the status quo maintained, and individuals settling down in their territories or their nests with their social status, their "mateships," and their lives more or less intact.

In the animal world—and possibly the human one as well (more on this later)— threats often involve bluffing, essentially claiming to be more formidable and more inclined to back up one's threats than is actually the case.[i] Generations of animal behaviorists have studied the various threat postures and vocalizations of different animal species. During the early days of so-called classical ethology as pioneered by biologists Lorenz and Niko Tinbergen, an optimistic and naive view of communication held sway. Sender and receiver were thought to be "on the same page," with senders indicating their internal state (aggressive, defensive, sexually aroused, and so on) by various postures and behavior, and receivers concerned only with decoding the messages as accurately as possible.

The messages themselves were assumed to be "honest." It was therefore believed that receivers could only gain by understanding and then acting on the information contained in such messages. Similarly, it was assumed that senders were well served by being the bearers of true tidings. This perspective has been challenged and, to a large extent, supplanted by a more cynical but probably more accurate viewpoint, deriving from the evolution-based assumption that living things (more precisely, their genes) are selected to maximize *their own* reproductive success and that, accordingly, animals aren't necessarily motivated to convey accurate information so much as to influence the behavior of recipients to enhance the fitness of the senders. From a hardline evolutionary perspective, maximizing the success with which genes get themselves projected into the future is not the primary focus of living things; it is the only focus.

---

[i] Among domestic cats, this involves literally getting one's back up (as well as piloerection—fur standing on end), both of which make the kitty look larger.

If the goal were simply to manipulate an inanimate object, animals have no choice but to rely on physical force: mostly pushing or pulling, or—in rare cases—using a tool. But if the object is, instead, another animal (e.g., keeping it out of your territory or away from your food, nest, or mate), then it is often far more efficient, as well as safer, to do so by hijacking the other animal's inclinations for the sender's benefit. Be sufficiently scary and your threat can essentially do the work for you.

Some animals send threats that are intended to be defensive, although human observers—feeling threatened—often perceive them as offensive and frightening. For Americans, the iconic example is the rattlesnake, which assuredly doesn't give itself away by rattling while hunting or before striking its prey. That distinctive sound, which people see as so threatening, is reserved for when the animal feels *itself* threatened.

Most often when someone is bitten by a rattlesnake, it is because the victim has accidentally stepped on the animal, who didn't get a chance to issue its self-protective warning. Normally, however, the potency of the threat—combined with its fangs and venom—is such that it works (warding off large-hoofed animals as well as against cougars, bobcats, coyotes, and wolves). The so-called Gadsden Flag, designed during the American Revolutionary War, made use of its emotional impact on people too: "Don't Tread on Me!"

Interestingly, this unique defensive threat has become disadvantageous for rattlers living in areas where they frequently encounter people who are, say, tending their garden. The snakes, feeling threatened, rattle to ward off a person, who, also feeling threatened, kills them with a shovel or rake. The unintended consequence of this shared threat sequence appears to be the evolution of silent, rattleless snakes, because quieter individuals have been more likely to survive and reproduce. This outcome, although presumably advantageous for those relatively reticent reptiles, isn't conducive to confidence among people liable to encounter snakes that are silent, but no less venomous. Rattlesnake rattles are honest threats; quiet rattlers are essentially dishonest in their silence, but in this case the lack of a communicable threat isn't their doing, but ours. And presumably, quiet rattlesnakes are at a disadvantage compared to their noisy colleagues when confronting another animal that, unlike some *Homo sapiens*, would just as soon keep away from the snake, if only it had been warned.

There do not seem to be any cases in the natural world—excluding encounters with human beings—of plants or animals being disadvantaged by their biological endowment of threats and weapons. This is because evolution generally operates with a kind of built-in regulatory mechanism, rather like a home thermostat, which turns the heat off when it gets too warm and on when it gets too cold. Should a species develop threatening weapons that are excessive and therefore disadvantageous to its possessors, natural selection works against such individuals, reproductively favoring those who are more moderate, and vice versa for those insufficiently endowed. For the most part, it is when people enter the scene that imbalances develop.

For example, blackberries are abundant in the Pacific Northwest, where, despite their luscious fruits, they are despised as weeds and often feared because of their aggressively protective thorns, structures that are generally effective in keeping away animals who would otherwise eat them. If blackberries were less threatening, home-owners would almost certainly cultivate them instead of eradicating them. Something similar (although without a deliciously compensating human benefit) pertains for poison ivy in the eastern United States, and for stinging wasps and hornets wherever they exist. When encountering human beings, threatening traits that otherwise serve their bearers well, and accordingly are favored by natural selection, often elide into liabilities. We shall see later that when it comes to people interacting with other people, or human social groups interacting with each other, this pattern of threats becoming disadvantageous is both persistent and widespread.

On the other hand, things that seem obvious and appropriate when we consider the animal world can be counterintuitive. The typical response of a threatened pit viper when confronted by a predator is first to flee, next to coil and emit a threat display, and then, finally—and only as a last resort—to strike. This sequence has led one herpetologist to say that "snakes are first of all cowards, then bluffers, and last of all warriors."[2] Research on cottonmouths—a highly venomous species—examined their two threat behaviors, mouth gaping and tail vibrating, subjecting the animals to a model human arm and observing whether a strike followed. Although gaping predictably preceded striking, tail vibrating did not.[3] Hence, the former—but not the latter—can be considered an honest threat signal in this species, and one that functions effectively.

It doesn't pay a potential victim to disregard warnings cavalierly, something that applies to threat behaviors in general and especially when dealing with such potentially serious outcomes as being bitten by a poisonous snake. This in turn opens the door for would-be threateners to exaggerate the danger that they pose, insofar as they (or rather, natural selection) can assume that their targets are unlikely to discount such messages.

Sometimes, as with rattling poisonous snakes, no exaggeration is needed, so long as the danger is sufficiently great. But even in such cases, clear and unambiguous communication is advantageous. There is, or was, an intriguing human parallel: pirates flying the skull-and-crossbones. Displaying the Jolly Roger was an effective way of issuing a threat, a highly credible one because pirates were known to be especially violent. Moreover, pirate credibility was also high because if captured, the consequences of announcing oneself—hanging—would likely be severe. As a result, only real pirates dared to fly the skull-and-crossbones. But the payoff was also large because by issuing a credible threat, pirates increased the likelihood that they could achieve their goal (whatever loot a merchant ship was carrying), without having to risk a fight. In June of 1720, for example, there were twenty-two merchant vessels in the harbor at Trepassey, Newfoundland when Batholomew Roberts, a renowned pirate, sailed in with the Jolly Roger flying. The merchant crews all panicked and abandoned their ships.[4]

There are very few reported cases of non-pirates using phony pirate flags. And there has only been one reported case of animals using the equivalent of the skull-and-crossbones: a physical structure, outside their bodies, to convey threat. Predatory birds known as black kites preferentially decorate their nests with pieces of white plastic. These avian Jolly Rogers are crafted only by black kites in prime condition, and they signal that their territories are comparably prime—and also that if another black kite dares to intrude, it is in for a serious confrontation.[5]

Mostly, animals species use their bodies to convey threats that are subtle but nevertheless consequential—and often quite dishonest. But first, let's look at honest threats directed particularly toward members of other species, and at the straightforward communication on which they are based. We'll then consider how some of these set the stage for dishonest threats.

WARNINGS AND MIMICS

Many animals are bad tasting or downright poisonous, not just in their fangs or stingers but in their very bodies. When they are caterpillars, monarch butterflies feed exclusively on milkweed plants, which contain potent chemical alkaloids that taste disgusting and cause severe digestive upset to animals—especially birds—that eat them, or just venture an incautious nibble. In the latter case, most birds with a bellyache are likely to avoid repeating their mistake although this requires, in turn, that monarchs be sufficiently distinct in their appearance that they carry an easily recognized warning sign. Not surprisingly, they are dramatically patterned in black and bright orange. To the human eye, they are quite lovely. To the eyes of a bird with a terrible taste in its mouth and a pain in its gut, that same conspicuous black and orange is memorable as well, recalling a meal that should not be repeated. It exemplifies "warning coloration," an easily recalled and highly visible reminder of something to avoid.[ii]

The technical term for such signals is "aposematic," derived by combining the roots for "apo" = away (as in apostate, someone who moves away from a particular belief system) and "sema" = signal (as in semaphore). Unpalatable or outright poisonous prey species that were less notable and thus easily forgotten will have achieved little benefit

---

[ii] It is no coincidence that school buses, ambulances, and fire trucks are also conspicuously colored, along with flashing red lights.

from their ostensibly protective physiology. And of course, edible animals that are easily recognized would be in even deeper trouble.

Warning coloration—bright and distinctive colors—emphasizes the threat, enabling a would-be predator to learn quickly who to avoid. Distinctive appearance itself thus becomes a form of threat: "Don't even think about taking a bite of me—or my genetic relatives—or you'll be sorry!" Many species of bees and wasps are aposematic, as are skunks: once nauseated, or stung, or subjected to stinky skunk spray, twice shy. But chemically based shyness isn't the only way to train a potential predator. Big teeth or sharp claws could do the trick, just by their appearance, without any augmentation. But when the threat isn't undeniably baked into an impressive organ—for example, when it is contained within an animal's otherwise invisible body chemistry—that's where a conspicuous, easy-to-remember appearance comes in.

Warning coloration isn't the only reason for notable and sometimes brilliant animal pigmentation. Sexual selection is responsible for much of the organic world's technicolor drama, such as the red of male cardinals, the tails of peacocks, or the rainbow rear ends of mandrill monkeys, all of which make these individuals more appealing to potential mates—probably because, once they are sexually attractive, they become more attractive yet because of what evolutionary biologists call the "sexy son hypothesis." This involves the implicit promise that females who mate with males who are thus adorned will likely produce sons who will inherit their father's flashy good looks and will therefore be attractive to the next generation, thereby ensuring that a female who makes such a choice will produce more grandchildren via her sexy sons.

Some of the world's most extraordinary painterly palettes are flaunted by neotropical amphibians known as "poison arrow frogs," so designated because their skin is so lethally imbued that indigenous human hunters use it to anoint their darts and arrow points. There is no reason, however, for the spectacular coloration of these frogs to serve only as a warning to potential frog-eating predators. As with other dramatically colored animals, colorfulness itself often helps attract mates, and not just by holding out the prospect of making sexy sons. Thus, in most cases, brightness is physiologically difficult to achieve, which means that dramatic coloration often indicates that such living billboards are also advertising their metabolic muscularity, indicating that they'd likely contain good genetic material. Moreover, it has been observed in at least one dramatically aposematic amphibian—the scrumptious-looking but highly toxic strawberry poison frog—that bright color does triple duty, not only warning off predators and helping acquire mates, but also signaling to other strawberry poison frogs that brighter and hence healthier individuals are more effective fighters.[6]

Once a direct threat (regardless of whether it has been combined with a biochemical boost) has been established, and whether or not the same trait also serves a romantic function, things get even more interesting. The honest threatener can become a model to be mimicked by other species that may not be dangerous to eat, but are mistaken for the real McCoy. Those monarch butterflies, endowed with poisonous, yucky-tasting alkaloids, are mimicked by another species—aptly known as "viceroys"—that bypass the metabolically expensive requirement of dealing with milkweed toxins while benefiting by taking advantage of the monarch's legitimately threatening reputation.

The plot thickens. Viceroy butterflies (the mimic) and monarchs (the model) can both be successful as long as the former are not too numerous. A problem arises, however, when viceroys become increasingly abundant, because the more viceroys, the more likely it is that predators will nibble on those harmless mimics rather than being educated by sampling mostly monarchs and therefore trained to avoid their black-and-orange pattern. As a result, the well-being of both monarchs and viceroys is diminished in direct proportion as the latter become abundant, which in turn induces selection of monarchs, which are discernibly different from their mimics so as not to be tarred with viceroys' innocuousness. But as the models flutter away from their mimics, the latter can be expected to pursue them, in an ongoing process of evolutionary tag, set in motion by the antipredator threat represented by the model's warning coloration, the momentum of which is maintained by the very different threats—to mimic and model alike—generated by the system itself.

This general phenomenon is known as "frequency-dependent selection," in which the evolutionary success of a biological type varies inversely with its abundance: greater when rare, diminishing as it becomes more frequent. (We shall see a well-studied example later in this chapter when we consider the so-called Hawk–Dove game.)

Warning coloration occurs in a wide range of living things, evolving pretty much whenever one species develops a deserved reputation for poisonousness, ferocity, or some other form of legitimate threat. Once established, it opens the door for what is sometimes labeled Batesian mimicry, because it was first described in detail by the nineteenth-century English naturalist Henry Walter Bates who researched butterflies in the Amazon rainforest. Brightly banded coral snakes (venomous) are also mimicked, albeit imperfectly, by some species of king snakes. Bees and wasps, with their intimidating stings, have in most cases evolved distinctive color patterns, often bands of black and yellow; they are mimicked in turn by a number of other insects that are outfitted with black and yellow bands but are stingless.

Interestingly, plenty of black-and-yellow-banded insects are in fact equipped with stings, although many other warning patterns are clearly available, and could use a variety of colors as well as alternative designs such as spots and blotches instead of bands. At work here is yet another evolution-based threat phenomenon, known as "Müllerian mimicry," after the German naturalist Fritz Müller. With this kind of mimicry, species that are legitimately threatening in their own right converge on the same pattern, the adaptive advantage being that shared appearance facilitates learning by predators: it is easier to learn to avoid one basic warning signal than a variety of them, different for each species. It had been thought that Batesian and Müllerian mimicry were opposites, with Batesian dishonest because the mimic is essentially a parasite of its model's legitimate reputation (those viceroys), whereas Müllerian exemplifies shared honesty (as with different species of wasps, bees, and hornets, whose fearsome reputations enhance each others').

It is currently acknowledged, however, that often the distinction is not absolute; within a given array of similar-looking Müllerian mimics, for example, not all species are equally honest when it comes to their underlying threat. The less dangerous representatives are therefore somewhat Batesian. Conversely, among some species, assemblages that have traditionally been thought to involve Batesian mimics—including the iconic monarch–viceroy duo—mimics are often a bit unpleasant in their own right, so both participants are to some degree Müllerian threateners.

DEIMATIC DEEDS

In addition to warning coloration and its hitchhiker phenomenon of mimicry, some prey species have a repertoire of threat behaviors explicitly directed toward their predators. In these cases, the prey hasn't merely been selected to remind the predator of its bad taste or venomous body fluids; rather, it mimics species that in turn prey upon those predators that would otherwise prey on the mimic. These dishonest or bluffing actions are often combined with anatomic structures that, in conjunction with suitable behavior, can startle a predator, especially when augmented by features that resemble the predator's predator. These anatomic structures, in conjunction with the behavior that calls attention to them, are sometimes called "deimatic," which comes from the Greek word "to frighten." For example, cuttlefish—small squid—often scare off predators by generating eyespots on various parts of their body that give the illusion of being large, scary fish. Significantly, this deimatic trick is only used when confronted by visually hunting predators. Others, which rely on sound or smell to obtain some calamari cuisine, induce the cuttlefish to simply swim away.[iii] Moths also deploy fake eye spots, but on their wings. These look like the eyes of owls, which are predators on birds that in turn would otherwise feast on the moths. Imagine a person about to eat a hot dog that suddenly resembles a poisonous snake. If a bird avoids a moth that suddenly looks like an owl, or recoils from a caterpillar that mimics a snake, these resemblances are likely deimatic insofar as the moth and caterpillar are actually edible—that is, if a

---

[iii] Cuttlefish are masters of deception, and not only when it comes to deimatic traits. For example, when a male is courting a female, he will typically adjust the color pattern of the side of his body facing his *enamorata* so as to maximize his romantic appeal, while simultaneously achieving camouflage on the other side, where he would otherwise be conspicuous to predators.

nonthreatening creature is giving a predator-mimicking threat display. On the other hand, if—after tasting the moth or caterpillar—the bird avoids similar ones because the creature really is foul tasting, mildly poisonous, or equivalent, then the display is honest.

Why don't Batesian mimics such as viceroy butterflies simply evolve their own honest threats, (e.g., toxic body fluids), to be announced by their own unique warning pattern? Probably because in most cases it is physiologically expensive to achieve the internal biochemistry needed to thrive despite carrying chemicals that are toxic to other creatures. It is less demanding to copy a model's color than its mechanism of detoxification. And also, as demonstrated by Müllerian mimicry, it is more effective to minimize the number of patterns a predator needs to learn—whether that learning occurs via experience or through natural selection.

Sometimes the world of communicated threats is remarkably Byzantine and unexpected. Renowned biologist J. B. S. Haldane—one of the twentieth-century's most creative evolutionary geneticists, and also a theoretician, physiologist, multilingual genius, and one-time activist in the UK's Communist Party—wrote that evolution is not only queerer than we suppose, but queerer than we *can* suppose.[iv]

Take the case of parasites manipulating their victims, which yields fascinating examples of "zombification." In most cases, when a parasite-ridden host is eaten, it is the end of the line for the parasite too. But sometimes parasites take over their hosts, inducing them to behave in ways that make the host and its hitch-hiking parasite more liable to be eaten by certain predators, which in turn facilitates the next stage in the parasite's life cycle. In at least one novel case, when insect larvae are infested by a particular nematode worm, the parasite induces an aposematic change in its host-victim, from its normally bland appearance to a vividly bioluminescent pinkish red. This transformation occurs even after the victim has been killed by the parasite, whereupon it serves as a visual deterrent to potential predators, giving them a false warning that the corpse is unpalatable, whereas in fact it is occupied by a parasite that would itself be killed if the dead host were eaten.[7]

TRUTH-TELLING?

Honesty, we like to think, is the best policy. In most cases, it probably is, and not just because as philosopher Immanuel Kant pointed out, truthfulness is desirable for its own sake, since a world of liars would be one in which the debasement of truth leads to a debasement of morality and thus a loss of social cohesion. There is also a more immediate and practical problem with dishonesty: you might get caught. This turns

iv I thank Richard Dawkins for correcting my initial misremembering of Haldane's actual words.

out to be especially cogent in the world of nonhuman creatures in which evolution by natural selection keeps score. Evolution is not concerned with theoretical questions of right and wrong, rather with what works and what doesn't—or, more precisely, what works better than its available alternatives.

Here, the conundrum comes not from ethics or even the difficulty of keeping track of deception in all its complex details("Oh what a tangled web we weave, When first we practice to deceive"[v]). The greater problem is that natural-born liars, in the biological world, must confront a practical dilemma often shared by their human counterparts: deceivers run the risk of being tripped up in their dishonesty, when and if their bluff is called.

A good reason for telling the truth is that you don't have to remember your lies, which, by definition, don't correspond to one's generally clearer memory of actual events. For animals, the issue is not so much remembering per se (although we'll see that reputation—especially what others remember about you—can itself be consequential), but having to live up to your announced capability. The story goes that when a gunslinger in the Wild West proved himself faster than the proclaimed "fastest gun," he was then in real trouble, because other fast guns would literally be gunning for him. In such cases, the short-term top gun must have been really good at his specialty or he wouldn't have ascended to such a precarious station. But what if someone simply proclaimed that he was really, really fast—but wasn't? That's even worse.

Nonetheless, dishonesty is widespread among animals, especially when it comes to threats. An empty threat can be dangerous for the threateners—more so than an empty promise—because although the latter can cause a loss of reputation, the former can lead to a fight in which the threateners, if their claim is "truly" phony, are liable to end up far worse off than if a hollow threat wasn't issued in the first place. George Bernard Shaw famously quipped that those who can, do; those who can't, teach.

Those that have effective weapons or other traits that can intimidate their rivals are not shy about doing so, by calling attention to those characteristics. Eastern collared lizards, for example, have strong jaws with which they can inflict serious gashes on an opponent during a fight, and even break bones. Males have especially powerful jaws, as indicated by the size of their masseter muscles. The threat display used by these animals involves an open-mouth posture in which the development of their jaw muscles is made clear. This display is geared to be seen, allowing—actually, more like forcing—a rival to recognize the fighting capacity of the threatening individual.[8]

Threat displays aren't always so honest, however. They often involve making the threatener's body appear larger than it actually is—for example, the way cats erect their

[v] This lament—and warning—often misattributed to Shakespeare, actually comes from Sir Walter Scott's poem "Marmion."

fur so as to appear as large as possible. American elk and Scottish red deer rely on their large body size and massive antlers to get their way, especially when it comes to competition between bulls. They frequently engage in "parallel walking," during which opponents strut their bodily stuff—notably, their impressive headgear. When successful, such demonstrations result in the more physically intimidating rival winning a contest without having to fight.[9] Predatory animals with large canines are not shy about displaying them toward an opponent. Poisonous snakes with impressive fangs not only reveal their weapons as needed, but emphasize them in the most intimidating way they can. Bighorn sheep call attention to their big horns. If you've got it, flaunt it.

It would seem that those who can back up their threats with genuine capability should go ahead and threaten, whereas those who can't would do well to maintain a meeker but more honest profile. Why, then, is fakery so widespread when it comes to animal threats?

To answer this we must first explore honesty versus dishonesty in animal communication more generally. An old tradition assumed that, when communicating, animals necessarily seek to maximize the accuracy of information transfer. The sender endeavors to convey certain persisting facts, such as its sex, state of health, and so forth, along with its current and potentially transitory state (sexually aroused, angry, happy, relaxed, and so forth), whereas the receiver tries to decode the message, which is assumed to be accurate. In short, sender and receiver were thought to be on the same page. The only disconnect was the physical space between them as well as the predictable psychological gulf that separates distinct individuals. The goal of communication is, therefore, to transcend this separation for the mutual benefit of both sender and receiver.

This cheery perspective has largely been rejected by today's biologists. The most widespread current interpretation of the evolutionary process is that natural selection focuses on genes, with success measured by effectiveness in projecting one's own genes or identical copies of these genes, lodged in the bodies of others, into the future. In short, evolution is persistently amoral, caring only for genetic success—not how that success is achieved—indifferent to the reproduction of genes present in other, unrelated individuals insofar as their DNA doesn't overlap.

Accordingly, when should particular gene carriers concern themselves with providing accurate or honest information? Only if and when that information will likely benefit the sender, and without regard to whether it is accurate or in any sense "true." It is a dispiriting perception, but one that seems in accord with how the biological world operates. There's another way of looking at this. From the perspective of an animal sending a message, the premium is on self-benefit, not honesty.

Moreover, communication is unlikely to be valued for its own sake and not necessarily in proportion as it tells the truth—about how things are in the outside world (nearby food, predators, the weather, and so forth), in the social sphere (who is in, who is out, up, down, and so on), or with regard to the communicator's particular situation (moods, desires, or other circumstances). This view may appear cynical, but it is nonetheless scientifically valid that, from a strictly evolutionary perspective, communication isn't so much information transfer as it is manipulation: the sender's effort to induce the receiver to do or to refrain from doing something, the goal of which is to enhance the fitness of the sender—more precisely, the sender's genes.[10]

This, in turn, puts a premium on the receiver's ability to figure out the message, which does not mean simply to interpret its meaning, but—equally important, if not more so—to ascertain whether acting on the information is beneficial to the receiver or whether, instead, it should be disregarded as a self-serving effort by the sender. In Shakespeare's *Henry IV, Part 1*, the mystical, egotistical Welsh warrior Owen Glendower brags, "I can call spirits out of the vasty deep," to which the bluff Englishman Hotspur replies "Why, so can I, or so can any man, But will they come when you do call for them?"

Why should those spirits (receiving the message) come when Glendower calls? According to evolutionary theory, only if the spirits determine that doing so is in their interest, because Glendower might otherwise punish them, because coming when he calls is somehow going to benefit them, or maybe because the unfortunate spirits have been somehow misled. However you slice it, Glendower—like all senders—is predisposed to manipulate receivers for the selfish benefit of his own genes, while those "vasty"-deep-inhabiting spirits, like all receivers, are equally predisposed to be fussy about their responses, coming only when the call is compelling *for them*.

So let's look at what might make a message compelling for the receiver. In all cases, it would have to involve a benefit to be achieved by the receiver, or at least perceived by the receiver to be in its interest. There are a number of scenarios in which the spirits might profit by heeding Glendower's call. Perhaps he and the spirits are genetic relatives, in which case it is not unlikely that, by inducing him to call, genes within the sender are benefiting identical genes within the spirits. This could include warning them about a predator, informing them of an opportunity to share food, and so forth. Or maybe by calling Glendower is reciprocating some benefit those spirits provided him in the past, or hopes to establish a mutually beneficial relationship for the future. Alternatively, maybe he and the spirits are part of a social network that requires a minimum number of well-provided individuals such that all of them are better off when any of them is aided. Also possible: Glendower is demonstrating his altruism to a third-party observer (Hotspur, in this case). In all these situations—and others can be identified—the spirits would be well advised to accept the sender's message pretty much at face value, just as, in all these cases, the sender, too, would benefit if the receiver acts on the message. That's why the sender is sending: not simply because he enjoys calling or is going out of his way to aid the recipients. These circumstances would all lead to more or less honest signaling, and thus, to more or less willing responsiveness by the receiver.

Alas, the biological world is not this benevolent and smooth functioning. There are many other reasons why Glendower might call the spirits, each of which would benefit him but not them. For example, maybe he is planning to eat them. (There is a species of firefly in which females mimic the mating flashes used by females of a different species to attract mates. When duped males respond by approaching, the mimetic females devour them.) Or perhaps Glendower sees some food and wants to divert the attention of the spirits so he can get it for himself. Similarly, if he has just spotted a predator, he might sound an alarm call not because he is generous with the information, but is trying to induce the recipients to move so the predator focuses on them instead of on himself. Or maybe he is trying to fatigue or distract the spirits so he can steal their stuff or their mates, or intrude on their territory. Or he could be showing off for Hotspur, demonstrating his power of persuasion, but just for his benefit, spirits be damned. In all these eventualities, plus others, it would behoove the spirits to look before they leap or, rather, to consider seriously what's in it for them before they accede to Glendower's call.[vi]

The resulting information arms race between sender and receiver plays out in many real-world contexts. When it comes to mating, for example, individuals of either sex can be predisposed to exaggerate their value as a sexual partner. This is especially true of males, because they can often increase their fitness by convincing more than one female to mate with them. A courting male in particular can then be faced with an especially acute need to convince a would-be partner that he is a good reproductive bet, because he is competing with other males who are selected to be similarly persuasive. Easier said than done, in part because, in many species, females—as the sex providing the resource that is especially valuable (her nutritionally rich eggs, which dwarf the metabolic value of a male's sperm)—are inclined to be especially choosy.

Equally demanding is the fact that talk is cheap; there is little to prevent a horny male from proclaiming his worth as a mating partner, which in turn selects for female ability to see through the male's efforts at self-promotion. And so, no matter how strongly he may be tempted to send false and manipulative signals, a courting male is faced with pressure to ensure his message is at least somewhat honest (or, at minimum, does an

---

[vi] Don't fall for the bogus criticism that in such cases living things cannot be expected to engage in detailed cognitive assessment, weighing the pros and cons of responding. Although it is certainly possible that high-order evaluations occur in certain circumstances, the detailed analysis is far more likely to have been conducted by many generations of natural selection and involve no more intellectual insight than is shown when a hungry squirrel eats an acorn without necessarily understanding the digestive and metabolic details of how doing so meets its nutritive needs.

honest job of seeming honest!), so as to breach his would-be mate's skepticism. This, too, is a challenge, one that can be met by engaging in courtship activities—or even by sporting physical characteristics—that cannot be faked, that can only be undertaken by individuals who are in fact of genuine quality. The likely solution to this conundrum was provided by Israeli zoologist Amotz Zahavi, who pointed out that the most effective courtship signals are those that contain a "reliability component."[11]

This can take the form of an anatomic impediment, such as the unwieldy and metabolically expensive tail of peacocks, the bright colors and often elaborate courtship antics in which many animals engage, and that to some extent have long been a puzzle to many evolutionary-oriented students of animal behavior. Big, awkward tails; conspicuous colors; and loud courtship songs often constitute a genuine handicap, absorbing precious calories to construct, making their possessors more vulnerable to predators, and taking up valuable time and energy. But it seems to be precisely these drawbacks that make the individuals who grow them and who thrive despite their liabilities especially attractive to females. This seems paradoxical, until we consider that such impediments proclaim to discriminating females that anyone so effective must be a formidable representative of their sex; otherwise, they wouldn't be able to manage their lives so well despite them. In short, they must have terrific genes. Anyone seeking to impress another might alternatively proclaim how wonderful they are, but such proclamations might be sheer braggadocio; being allied to an unmistakable liability ensures the reliability of their message—a guarantee that something about them is as proclaimed.

A similar reliability component operates when it comes to some otherwise perplexing behavior of prey species. Prey rarely have much to communicate to their predators, aside from such dishonest messages as "I'm not here" (camouflage) or "I'm really just a dead leaf or a bit of bird poop" (mimicry), or "I'm bad-tasting, poisonous, or otherwise dangerous (Batesian or Müllerian mimicry). Most of the time, and for understandable reasons, predator and prey each seek to remain undetected. In some cases, however—notably Thomson's gazelles in Africa and, to a lesser extent, white-tailed deer in North America—prey species do something counterintuitive when they spot a predator and when they know the predator has spotted them.

Gazelles in particular frequently jump high in the air, holding their legs oddly stiff and resembling an animated four-legged pogo stick. In doing so, they make themselves more conspicuous to the predator (especially cheetahs) and expend energy that might be used more adaptively to run away. But in fact, this action—known as "stotting"—is quite adaptive, sending these messages to even a hungry cheetah: for one, I have seen you, so you might as well give up chasing me because the element of surprise is foreclosed. And, for another, look how strong, well-coordinated, and agile I am! Stotting is a demanding bit of gymnastics, something that only a strong, well-coordinated, and agile gazelle can master.[12] The action is its own reliability component.

Only relatively young and healthy animals stot, and predators are unlikely to try their luck against individuals who do so. Presumably, gazelles could announce their physical fitness in other ways, but anyone—regardless of actual capability—might simply lie about it. As with birds who flaunt their fancy, expensive, inconvenient tails, stotting gazelles are proclaiming their personal qualities, and doing so in a manner that cannot be faked. When African wild dogs—formidable predators of gazelles—switch their pursuit from one individual to another, they are likely to change to a gazelle who has been stotting less vigorously, or not at all.[13] North American white-tailed deer do not stot like African gazelles, but they do something analogous: vigorous tail-flagging, which reveals the white rump patch and underside of the tail, which in turn makes them more conspicuous to predators, certainly not less. This signal is especially visible to cougars or wolves, from whom they are running away, rather than being a signal to other deer.[14] Moreover, they do it as much when alone as when in a herd with other white-tails,[15] further suggesting that tail-flagging is a signal to predators, indicating "I've already seen you" and "Nya-nya; you can't catch me." So don't even try.

Nonhuman primates don't typically concern themselves with sending honest versus dishonest signals to predators, à la stotting in gazelles. But, several different monkey species in the Taï Forest of West Africa have to deal with hungry leopards and chimpanzees. These two predators hunt monkeys very differently: chimps by persistent pursuit and leopards by sudden surprise attack. When Taï Forest monkeys perceive a leopard, they are likely to give an alarm call, saying, in effect, "I know you're there. You can't surprise me." But because they don't rely on surprise, chimpanzees aren't deterred by a message indicating they have been seen,[16] and accordingly, chimps don't evoke a comparable response from their potential prey.

Credibility doesn't only count when it comes to impressing a romantic partner or a predator; it is especially key when communicating threat. Here, the temptation to send dishonest messages is particularly strong, because threatening animals are trying explicitly to influence the behavior of the recipients, typically in circumstances of high tension: getting them to refrain from entering their territory, courting their mate, or trying to horn in on some food. The intent of the threat is often to convince the recipient to back down and decide not to challenge the threatener's social position. This is easy enough if you are genuinely strong, fierce, well-coordinated, highly motivated, physically fit, and so on, but what if you're less than imposing? Or if you are actually quite a tasty prey item but would be more successful evolutionarily if you could convince your predators you are actually distasteful or, better yet, downright poisonous? Maybe then, *dishonesty* is the best policy—unless you get caught in the lie.

In view of the high payoff that attaches to success, there must be a substantial evolutionary temptation to make empty threats—that is, to exaggerate one's ability to back up a threat with physical action. Doing so, however, can carry a substantial cost if the threatener's bluff is called and it ends up in an altercation that could have been avoided. There might also be a dangerously negative effect on one's reputation. Although human beings take comfort when a dog's (or a person's) bite is less than its bark, the cost of being branded a paper tiger or "biteless" barker can be quite serious. The temptation to make empty threats is leavened accordingly by the cost of being found out.

If signals are largely deceptive and manipulative, then receivers should evolve to ignore them and, as a result, signalers should eventually be selected to refrain from empty posturing. The overwhelming likelihood, therefore, is that some threats most of the time, or most threats some of the time, actually convey useful information. Amotz Zahavi was an ornithologist, although his idea applies across the animal world (and, as we'll see, some important parts of the human world too). As Zahavi put it in his now-classic article with regard to courtship, "Since good quality birds can take larger risks, it is not surprising that sexual displays in many cases evolved to proclaim quality by showing the amount of risk the bird can take and still survive."[17]

This applies equally to threat displays, which are also costly, because such actions, especially if prolonged, are not only exhausting, but liable to be conspicuous—not only to their intended target, but also to lurking predators. When it comes to issuing threats to a same-species competitor, however, most threat displays do not appear to involve a notable handicap, because threatening behavior generally involves deploying the same weapons that would be used if the threat is unsuccessful and a mano-a-mano, beak-to-beak, claw-to-claw, horn-to-horn, antler-to-antler, or mouth-to-mouth battle were to ensue.

The evolution of animal weaponry thus tends to incorporate reliability components into the structure of the weapons themselves—not so much because of the adaptive value of such signals, but because most weapons are easily visible and, moreover, the elaboration of large, impressive, and thus especially threatening weapons requires an investment that only strong, healthy individuals can make. Just as the reliability of gazelle stotting is inherent in the behavior, the guarantee against dishonesty provided by most animal weapons is built into the anatomy itself and cannot be faked easily.

Nearly all living things are under energetic constraints, resulting in what economists call "opportunity costs": an investment in one domain typically occurs at the cost of reduced opportunity to invest in another. This has been documented in dung beetles, among which the males develop horns that are used both to threaten and to fight each other. When generations of males were selected in a laboratory to have exceptionally large horns, the resulting animals ended up with stunted eyes, wings, genitalia, and so forth—a seemingly unavoidable trade-off that emphasizes the opportunity costs of evolving immensely large weapons.[18] Clear-cut trade-offs of this sort, however, seem limited to insects, among whom the development of various body organs—including weaponry—is initiated early in metamorphosis.

In contrast, among most other species, from crustaceans to elk, body parts are pretty much fully formed before weapons begin to develop. The requirements to build a massive crab claw or a rack of ungulate antlers are nonetheless quite demanding, requiring the animal to spend less metabolic resources on the rest of its body. Male fiddler crabs, for example, have the largest weapons per body size of any animal; fully one half of a male fiddler's energy budget is expended in growing its comparatively huge fiddles. Even after they have been produced, these structures impose a resting metabolic rate on their possessors that is twenty percent greater than that of females (which don't have enlarged claws), simply because of the cost of maintaining so much additional muscle.[19] And this doesn't count the expense of waving their appendages vigorously, as well as added demands when it comes to running with them—especially from predators, something that might well be particularly mandated for male crabs (and, to a lesser extent, lobsters), whose oversized claws make their possessors especially desirable prey items because of all that expensive muscle meat inside.

The costliness of animal weapons is not limited to invertebrates. The antlers of male caribou can weigh more than 20 pounds and be 5 feet or more in length; moose, 6.5 feet and 40 pounds. Balanced at the top of the animal's head and at the end of a long neck, the encumbering weight of these structures is increased by a levering effect. Beyond the stress of walking, running, displaying, and occasionally fighting with them, there is also the cost of growing these devices, estimated at five times the energy requirement of merely keeping the body going.[20] As if this isn't costly enough, the demands for calcium and phosphorus to produce massive antlers, for example, necessitate that a mature bull elk or moose literally depletes the mineral content of his bones, resulting in a kind of seasonal osteoporosis, causing bone brittleness and rendering males susceptible to life-threatening fractures.

It seems pretty clear that whatever their obvious functionality, major weapons impose a cost with an expense that makes them additionally credible insofar as only well-endowed individuals are able to afford them. And yet, an alternative possibility must be admitted: because they are so demanding to produce and deploy, such armamentaria necessarily deplete the body resources of individuals who manufacture them, so that large, hefty claws or massive, ungainly antlers might actually indicate weakness rather than strength, given that their carriers have endured a kind of metabolic deficit spending. The evidence, however, is otherwise: animals with expensive, overbuilt weapons are nearly always treated as formidable opponents rather than musclebound convalescents, worn out from their exertions to impress. Weapons, including seemingly excessive ones, thus appear to be highly reliable as implements of functionality, no less than of threat. (At least among animals.)

Reliable signals are especially useful, for both sender and receiver. Insofar as such signals are credible and thus taken seriously, senders would be less liable to endure challenges, and receivers would be less inclined to waste time and energy testing the legitimacy of the message. Moreover, when it comes to threat behavior, signals don't necessarily have to be attached to a handicap. Intuition would suggest, in fact, that threat signals that are both reliable and handicap free would be especially prized. For a peculiar example, consider scent-marking by female dwarf mongooses. In this species, females are the main competitive threat to other females. They indicate their presence and issue a challenge by squeezing the products of their anal glands on rocks and especially vegetation, and they do so in a way that human observers cannot help finding comical: by performing a handstand while thrusting their hind end to maximum elevation.

This bit of gymnastic self-advertisement is a reliable indicator of body size, because in proportion—as they aren't especially dwarfish—nominally dwarf mongooses are reliably able to squirt their stuff higher.[21] Even when their own scent mark is about the same height as that deposited by an earlier marking mongoose, these animals have a competitive maneuver available: they "overmark," spraying or smearing their secretions on top of their predecessors'. Bizarre as the physical act looks, handstand scent marking—especially urination—is also used by other mammals, including female bush dogs[22] and male giant pandas.[23]

As animal threats go, scent marking is among the least risky, because it is done *in absentia*: the scent purveyor is typically not in the immediate vicinity when it is received. It is also largely handicap free, although given that it is extremely difficult for vertically challenged individuals to fake the altitude of their scent mark, this particular technique is imbued with its own reliability component. As noted, the most common animal threats also carry a substantial reliability component insofar as they involve a display of structures that would also be used if push comes to shove.

For the same reason, they generally lack an immediate associated handicap because there is nothing inherently risky in calling attention to one's own formidable anatomy, which is not rendered less usable by being highlighted. Hence, it isn't necessarily so that, to be reliable, a threat signal must put the signaler at a disadvantage. It could be argued that, in the case of aposematic coloration such as those lovely poison arrow frogs bright colors and thus, greater conspicuousness renders their possessors more vulnerable to predation and thereby constitute a potential handicap. But, because those bright colors also enhance their conspicuousness as a threat, the issue seems moot.

Later we encounter the question of reliability in its most dangerous and subverting manifestation when we examine the problem of credibility in the realm of nuclear threats—namely, deterrence. For now, let's note that reliability haunts much of interpersonal human behavior more generally, with empty threats and promises being both tempting and risky. As with most animal species that issue threats, the payoff to successful bluffing can be high, but so is the cost of being found out. Braggards are not admired, although they can also benefit from their braggadocio. Case in point: Donald Trump, who never seems to miss an opportunity to enlarge upon his purported brilliance, accomplishments, and the extent to which he claims to be admired by others. And enlarge he does, trumpeting his imagined achievements—along with threats to those who cross him—with such seeming confidence that many people among his "base" are taken in, as evidenced by his election to the US presidency.

On the other hand, perhaps the most important enforcer of honesty in our own species is the negative consequences for one's reputation if personal reliability is lost. *Homo sapiens* have, accordingly, elaborated numerous mechanisms for ensuring reliability or, at least, encouraging it. We make oaths, often invoking God and/or one's ancestors, solemnly swearing to speak the truth, the whole truth, and nothing but the truth. We also establish legal consequences for perjury. In ancient Rome, oath-breaking was a capital crime, with perpetrators thrown off the Tarpeian Rock—although, according to Cicero, when it came to oath-breaking, the primary punishment was carried out by the god in whose name the oath was originally made, with the immediate, human-imposed consequence being loss of reputation. Diminished reputation is typically an immense liability among most human beings, especially because, unlike other species, we often rely upon our "word of honor" rather than such physical reliability components as an expensive feathery tail á la peacocks or a massive rack of antlers as in bull elk.

During the Cuban Missile Crisis, President John F. Kennedy was eager to obtain the support of his European allies. Former Secretary of State Dean Acheson was accordingly dispatched to Paris, where he informed President Charles de Gaulle that the Soviet Union had been emplacing medium-range nuclear missiles in Cuba, and he offered to show photographic evidence. That was not necessary, de Gaulle replied; the word of the President of the United States was more than sufficient. National leaders have generally endeavored to maintain a reputation for reliable truth-telling, for precisely this kind of situation, when a crisis puts a high premium on credibility.

Mr. Trump, in contrast, has been notoriously indifferent to the truth, regularly flouting the supposed need for honesty in communication, whatever the species. According to *The Washington Post*'s Fact Checker (a full-time job since Mr. Trump became president), he had committed 16,200 incidents of "exaggerated numbers, unwarranted boasting and outright falsehoods" by January 20, 2020, after three years in office, averaging six a day in 2017, increasing to 16 per day in 2018, and then topping more than 22 a day in 2019.[24]

It is most unlikely that any modern-day national leader would have responded to Donald Trump as President de Gaulle did to the emissary of President Kennedy. Because he has persistently ignored the basic bio-logic of reliability as the key to honest communication, President Trump—and, by extension, the United States—has achieved less credibility than an eastern collared lizard or a stotting gazelle.

HAWKS, DOVES, AND OTHER GAMESTERS

The elaborate variations of animal threat have been studied extensively by biologists and also subjected to intensive computer modeling. The best known of these is the Hawk–Dove model, which lends itself to analysis via mathematical game theory. Its basic idea can, however, be conveyed without equations or (game theoreticians' favorite medium) payoff matrices. Here goes. Imagine two types of animals. Hawks are aggressive competitors, who threaten whomever they meet and are prepared to escalate the confrontation into a fight if need be. These one-to-one encounters are assumed to take place over a mutually desired resource, such as a mate, nest site, food item, and so forth. Doves are peaceful, inclined to give way when confronted by a Hawk. Accordingly, when Hawk meets Dove, the Hawk makes Hawkish threats and gets the resource in question because the Dove simply gives up (score one for the Hawk), but at least the Dove departs uninjured and without having spent any time engaging with the Hawk. When Dove meets Dove, they neither threaten nor fight; instead, they divide the resource amicably, each getting one half. When Hawk meets Hawk, both threaten, escalate the confrontation, and eventually fight. They also divide the resource (as do Doves when meeting another Dove), but only after enduring the cost of battle, a downside from which Doves are exempt.

Imagine, now, a starting population composed exclusively of Doves, who regularly encounter each other. Each does fine, suffering no great loss and regularly getting its share of the contested resource until Hawks arrive, either via immigration or mutation. Each Hawk will initially have a field day because it meets Doves and inevitably gets the resource while the Doves are left with nothing. As a result, Hawks—initially rare—will increase in the population because they are getting more than their share of resources, which are defined as anything that enhances reproductive success.

But this success carries within it the seeds of their undoing, because as they become increasingly abundant. Hawks begin running into other Hawks, and when they do, the result is a debilitating fight. If the cost of such fighting is greater than the value of the resource they obtain postconflict, then Hawks will do poorly compared to the Doves they had previously displaced because, by definition, Doves don't fight and are spared the disadvantage of such confrontations. Accordingly, Doves will then increase in abundance relative to Hawks, until they become so frequent they are again vulnerable to invasion by newly arriving Hawks, and the cycle begins once more. Depending on various conditions—notably, the value of the disputed resource relative to the cost of threatening and then fighting—the ultimate outcome can be an equilibrium involving various proportions of Hawks and Doves, or a system of continuous pendulum swings.[25]

Other models, more complex and presumably more realistic, have also been developed. Some of them assume the participants can modify their actions depending on the behavior of their opponent. Let's therefore add to the mix a strategy known as "Bully." When they meet Doves, Bullies act like Hawks, threatening and pretending they are ready to fight. This benefits the Bullies because the Doves run away. But, when confronted by a Hawk, Bullies reveal their inner coward and switch to dovishness. It turns out that bullying is generally a successful strategy, because it does well against Doves (by acting like a Hawk) but avoids mutual escalation with Hawks (by acting dovish). When two Bullies meet, a likely outcome is that each acts like a Hawk half the time, whereupon the other acts like a Dove and the former gets the resource. Equally often, the payoffs are reversed; but, in either case, hurtful escalations are again minimized.

Next, add another behavior type: call them "etaliators." They are in some ways the inverse of Bullies, behaving as Doves at the onset of any meeting, so when they encounter a genuine Dove, everyone shares the resource nonconfrontationally. But if the other individual threatens and then escalates (that is, reveals itself to be a Hawk), Retaliators also act like Hawks.

Retaliators do the best of all, because when they meet a Bully—who escalates to hawkishness, mistaking Retaliators for Doves, they once more escalate to hawkishness themselves, whereupon the Bully runs away. It is interesting that, in nature, many animals are Retaliators, escalating only in response to another's escalation. A defeated rhesus monkey, for example, will tolerate minor damage inflicted by the victor's incisors, but if the latter escalates and uses its more formidable canines, a retaliation usually ensues. In short, a defeated individual typically acts like a Dove; but, if the victor goes too far, the Retaliator within is revealed.

When it comes to aggressive threats, the issue isn't only one's inclinations. There is also the question of the threatener's actual fighting ability—the extent to which it is physically able to back up its claims—along with its inclination to do so: the probability that it will use whatever ability it possesses. This is analogous to what military analysts describe as "capabilities" versus "intentions." The former is easier to assess than the latter. For example, how long or sharp are its canines? How large is the animal's body? How many soldiers or tanks does a country have? Such assessments are relatively straight-forward. More difficult is whether the other side will use its military forces, or whether another animal will use its weaponry. Is one's rival a Ferdinand among bulls or a paper tiger among nations? A Dove, a Bully, a Hawk, or a Retaliator? (And this is an oversimplified case of merely four options.)

In addition, a given resource—whatever it is that the contestants are contesting—is nearly always valued differently by the players involved. The one valuing it more highly is more likely to threaten more—to be more hawkish—or, at least, to pretend more hawkishness, risking confrontation in hope of getting its way.

Whenever two parties are engaged in a contest and each has a limited palette of possible behavior, and if two are threatening each other and are equally matched in strength and fighting ability (i.e., in capability), the one that is more inclined to escalate (aggressive intention)—because it is hungrier, hornier, more concerned about its reputation, or otherwise more strongly motivated—will win. So far, honesty is rewarded, insofar as the more driven individual is genuinely more aggressive and receives a positive payoff as a result. In fact, it would be adaptive for both if each were to signal its degree of commitment honestly, which would resolve things without a fight.

"You really, really want to win? Because I'm more indifferent. OK then. No big deal. It's yours."

Or, similarly, "You really are an impressive specimen! All right, I've got other things to do."

But, such a system could, in theory, be invaded by liars who proclaim a desperate passion, and thus willingness, to escalate beyond their actual ability or inclination or who convey an inflated impression of size, strength, or some other indicator of their capabilities. These liars would do well initially, because they can win by bluff alone, without having to fight. Their number would therefore increase, and empty threats would become the norm.

But, as in the basic Hawk–Dove game, in which Doves are vulnerable to invasion by Hawks, eventually a population of mostly cheaters would likely be tested by individuals willing and able to call their bluff and to back up their threats. (For individual animals, we can once again substitute countries.) But then, after the honest testers increase in number and—as a result of their abundance—find themselves confronting other honest testers, the resulting tests would likely be testy indeed, resulting in potentially injurious fights, equivalent to Hawk–Hawk interactions.

At this point, once again analogous to the situation of Hawks as hawkishness becomes the norm, after escalating fights become sufficiently frequent, the system is ripe for invasion by those dishonest bluffers, who could benefit from an increasing tendency of testers to back off when challenged, resulting once more in a spike in the number of liars, which could reset the system again. Over time, the outcome could be either ongoing pendulum swings between mostly honest and mostly dishonest threateners or a stable equilibrium between the two strategies, depending—as ever—on the costs and benefits in each case. (Recall the monarch-viceroy situation, in which an excess of viceroys puts them at a disadvantage.)

Of particular interest for our purposes is that would-be liars can be kept honest by handicaps—reliability components—associated with the threats themselves,[26] if, as seems likely, threatening is costly. This assumes there are two kinds of individuals (strong and weak) along with two different kinds of signals (a fiercely threatening one that is given when the sender is strong, and another, less intimidating, given when weak).

Honest communication of threat then takes place when strongs act strong and weaks act weak. But as before in the case of dishonesty versus honesty, what is to stop weaks from pretending to be strong? They are likely to be successful against other weak individuals, but to have a problem if they encounter strongs, assuming the strongs are inclined to challenge deceptive signalers. Giving such a dishonest signal would then be costly for the weaker individual, so it would be a kind of handicap, something that falls more heavily on the less competent one. But unlike stotting, lying in this way is something that a less competent individual could nonetheless signal, but presumably would be hesitant to do. Hence, dishonesty itself can carry a kind of reliability component, or at least, a built-in cost of being too dishonest or of being one of too many who are somewhat dishonest.

There is some reality undergirding all this theorizing. In most bird species, the intensity and brightness of feathers correlate with physical condition, and not surprisingly individuals who display such conspicuous badges achieve social dominance and are accorded the benefits thereof: better nest sites, more mating options, priority at contested food sources, and so forth.

Among white-crowned sparrows, a common species found in North America, adult males sport striking black-and-white stripes, whereas the coloration of juveniles and females is more subdued, with a pattern of alternating light and dark tan. When the badges of young females were enhanced experimentally by enamel, they ended up being dominant over controls.[27] This, in turn, raises the question of why some individuals settle for a less impressive appearance, given that the more dramatic the color contrast sported by an individual, the greater the social deference it enjoys, with the reproductive benefits that conveys.

One answer may be suggested by research on Harris sparrows, a closely related species. These birds also possess badges that are associated with social dominance—in this case, the darkness of their breast feathers. When first-year males and females—whose feathers are pale—were blackened with hair dye and then released into a free-living flock, their dominance status increased.[28] However, they quickly experienced significantly more attacks from dominant males than they had received pretreatment.[29] In this species (and others as well) aggression is particularly likely between individuals who sport comparable badge sizes and intensity—which makes it especially risky for subordinate, small badge-bedecked individuals to fake their capability and assume a threatening appearance to which they are not "entitled," by producing a badge that is darker and bigger than his britches.

This likely explains why individual males don't grow larger and darker badges than they deserve. With white-crowned sparrows, in contrast, there doesn't seem to be a comparable cost to cheating, and yet subordinates and young don't do so. Why not? One possibility is that it can be not only behaviorally risky (e.g., Harris's sparrows), but also metabolically expensive (white-crowned sparrows) to be a Beau Brummel among birds. If so, then not all individuals can flaunt their healthy vigor by dressing up in fancy feathered attire—even if they might prefer to do so.

SHRIMP SALAD

The most convincing examples of honest versus dishonest threats among animals comes from certain crustaceans: aka snapping shrimp and mantis shrimp. Pound for pound—more accurately, gram for gram—these are among the most ferociously armed of all animals. Their "raptorial appendages" consist of either weaponized clubs that have immense striking power, capable of breaking mollusk shells, or strong and sharp forelimbs used to catch and pierce prey. The former species is sometimes known as "smashers" and the latter, "snappers."[30] Either weapon type can readily injure or even kill same-species opponents.[31] Not surprisingly, their weapons are brandished conspicuously as parts of aggressive threats in a way reminiscent of the more widely known horns of mountain sheep[32] or the antlers of Scottish red deer[33] (equivalent to what are called "elk" in North America[vii]).

Compared to these massive mammals, shrimp are indeed shrimps; but, for their size, they are—if anything—more dangerously accoutered. The largest of these raptorial appendages, called "meri," are part of a male's weaponry and are extended and directed toward other males as part of an aggressive signal called a "meral spread."[34] These displays are especially pronounced and important when the animals are competing for

[vii] Confusingly,]what the British call "red deer" (Cervus elaphus) are very closely related to *Cervus canadensis*, which are known as "elk" in North America, while what Americans and Canadians call "elk" are often labeled "moose" in the United Kingdom. So far as I'm concerned, "moose" (*Alces alces*) are moose, period, although what North Americans call "moose" are sometimes called "elk" in the UK and Europe. Confused? So am I! All the more reason for the Latin taxonomy, although sticking with plain English, one way out of this confusion is to identify North American elk by their native American name, wapiti, just as it is helpful to distinguish North American "buffalo" from their Old World counterparts by designating the former as "bison." Perhaps some day—albeit not soon—Europeans and Americans will agree to use one word for caribou and reindeer, which are the same critters.

a particularly cherished resource: dwelling cavities in rock or coral. Meral spreads by mantis shrimp make the threatening individual less liable to be attacked and increase the probability that an intruder will back off and stop intruding.[35]

Another species, snapping shrimp (also called "pistol shrimp"), use a different threat behavior—the "open-chela display"—whereby their very large claws (chelae) are held open like a cocked pistol and waived at an opponent. When a researcher collected molted chelae (empty claws, without an attached animal) and presented these in an open position to experimental subjects, the response was dramatic: they responded by displaying in direct proportion to the ratio of their claw size to that of the claws to which they were exposed. That is, intact animals with larger chela compared to the test object engaged in more displays, whereas those whose chelae were smaller than those of the test objects did relatively less displaying.[36] This is to be expected, given that chela size is a good proxy for body size and fighting prowess, with larger individuals seen as more threatening because they are more likely to win.[37]

The meral spread display generally indicates threatening intent on the part of mantis shrimp; a targeted individual is less likely to attack a displaying animal. But here is where things get especially interesting. These crustaceans—collectively known as "stomatopods"—molt regularly, emerging from their discarded carapace. Immediately after doing so, their bodies are soft and vulnerable, so they cannot withstand even light blows from an opponent. Moreover, in their unprotected state, they are also unable to use their chelae to initiate an effective attack. In fact, the odd times when a soft-bodied, freshly molted individual attempts to strike with these weapons, it injures itself instead of its opponent.[38] Nonetheless, they still display. Because a freshly molted individual cannot effectively back up his display with an actual attack, it's all bluff.

In an especially clever experimental setup, newly molted and thus vulnerable individuals were given ownership of highly desired rock cavities in the laboratory. Hard-shelled individuals were then introduced. They sought access to the already-occupied cavities. In most cases, the newly molted residents promptly gave up their real estate, although some first engaged in threat display (i.e., they tried bluffing before retreating). Controls—hard-shelled cavity owners who were between molts and thus as physically competent as the intruders—had no need to bluff. They responded to trespassers by physically attacking them (fifteen out of nineteen pairings), with just four giving a meral display only. In contrast, among those helpless, newly molted cavity residents who resisted the newcomers, the proportions were almost exactly reversed; two lunged aggressively at the intruders, whereas fifteen used a meral spread threat display but then promptly gave up when it was unavailing.[39] Reworking George Bernard Shaw's quip about teachers versus doers, those that can actually defend their turf do so. Those that can't, act as though they can. In other words, they threaten. And then leave.

As W.C. Fields is alleged to have recommended, "If at first you don't succeed, try, try again. If you still don't succeed, give up. No sense being a damned fool about it."

In a follow-up study, researchers varied the size of intruders versus resident owners, whereupon the frequency of threat displays by recently molted residents increased in proportion as their body size exceeded that of the less vulnerable intermolt intruders. That is, as residents appeared more formidable (even though they weren't—remember, these were recently molted and therefore soft-bodied individuals), they were more likely to threaten and less likely to flee. At least initially.

Their fakery increased in proportion as they were able to give the impression of being more formidable because of a body size discrepancy, even though in reality they were no more able to back up their threats than were any other recently molted residents. On the other hand, among hard-shelled residents (less vulnerable and also enjoying the psychological advantage of being territory owners), those who were larger than the intruders gave proportionately fewer threat displays and proceeded directly to attack.[40]

Also worth noting: those residents who bluffed—who gave a threat display despite having recently molted—were more likely to retain their ownership than were those who refrained from displaying. In short, if you're a recently molted stomatopod shrimp, bluffing pays—at least much of the time, and as observed in the laboratory. Among free-living individuals, if a hard-shelled intermolt enters the cavity of a vulnerable, recently molted animal, the latter is often killed.

So, by threatening the intruder, vulnerable animals are taking a big risk, but also giving themselves a chance of winning. Their prospects are somewhat enhanced because, in natural populations, roughly eighty-five percent of cavity residents who give meral spread displays are hard-shelled intermolts, whose displays are therefore "honest."[41] This gives the bluffers a chance, because intruders who perceive these threatening messages would be well advised to believe they are backed legitimately. (Think back to those pirates of the Caribbean and elsewhere, who, by displaying their pirate flags, were honest in at least one respect: being the equivalent of hard-shelled cavity owners relying on their genuine, well-earned reputations.)

There is even reason to think that *individual* stomatopod shrimp vary their behavior in an effort to enhance their personal reputation for reliable aggressiveness, thereby increasing the likelihood that when they have recently molted—and are therefore vulnerable to attack as well as unable to back up their own aggressive threats—they will nonetheless be treated as though they are formidable and should not be challenged.[viii]

[viii] There don't seem to have been any pirates who, unable to fight effectively (perhaps because they were undermanned or out of ammunition), flew the skull-and-crossbones anyway and thereby succeeded in bluffing their way to success.

Individual shrimp are able to recognize each other by odor (especially those with whom they have had aggressive encounters), and they tend to avoid opponents who had previously defeated them; they also avoid the cavities maintained by victors. This makes it all the more useful for individuals to pair their threatening meral spreads with actual attacks while they are between molts, so as to enhance their reputation for those times when they will have a dangerously soft cuticle. To succeed at such times, they have to do so by reputation alone. In this regard, it is probably not coincidental that animals become more aggressive shortly *before* molting, and that they are also more likely to combine meral display threats with actual claw strikes at this time.[42]

Crustaceans in general—shrimp, crabs, lobsters, crayfish, and so on—use their claws to threaten each other, but also to obtain food. It turns out that, among crayfish and possibly other crustaceans, size matters—especially claw size—because they are used for both threatening and fighting. Interestingly, however, size does not predict pinching strength, so claw size is, to some degree, a dishonest signal of physical prowess. Males develop exceptionally large claws, used more for intimidation than for fighting or obtaining food. Moreover, muscle fibers from males produce only about one half the force as does comparable muscle mass from females. So males invest more in show; females, in the real thing.[43]

VOCAL THREATS

Neither crayfish nor shrimp appear to have a lot to say, although many animal threats are conveyed vocally. When this is the case, deeper sounds (lower frequency) are nearly always taken more seriously because they signal a more imposing opponent. Among many species of frogs and toads, deeper croaks indicate larger bodies and, thus, greater physical capacity. The key element in these calls is the "dominant frequency," the number of Hertz in which most of the calling energy is concentrated. In a common species of European toad—as in many other anurans—males clasp females during mating in a posture known as "amplexus." When a second male approaches a male engaged in amplexus, the mating male gives a call and, the deeper it is, the more likely it is that the intruder backs off.

This was demonstrated when researchers gave small males the opportunity to mate with receptive females and then introduced intruder males. The small males in amplexus would normally give a call characteristic of their size (higher pitched dominant frequency); but, regardless of how the intruders responded, this would not in itself demonstrate the role of vocalization as a way of warding off the intruders, because the diminutive calling males would also be visibly smaller. So, in an intriguing semicomic twist, the ingenious researchers silenced the mating males by hooking a rubber band just behind their front legs and through their mouths. Then, using a loudspeaker, they played vocalizations from either large or small males, whereas in either case the mating individuals were the same size. Intruders were three times more likely to attempt to supplant the artificially silenced mating males when a high-pitched call, associated with small individuals, was played.[44]

These basic results were confirmed in free-living animals as well, when low-frequency versus high-frequency calls were played to toads and frogs in their natural environments. In both cases, calling males were significantly more likely to swim away from deeper calls than from higher ones.[45] Among amphibians, the depth of calling frequency varies less with the length of the vocal cords than with their mass (actually, the size of a thick, fibrous structure generally located at the center of the vocal cords), which in turn varies with body size. Cheating in this regard is therefore possible, because even small-bodied individuals could presumably evolve deceptively beefy vocal cords.

There may nonetheless be a reliability component operating here, because producing such structures could be a metabolic expense that exceeds the energy budget of smaller animals. Alternatively, they could lower their calling frequency by reducing the pressure at which the cords vibrate, but this necessarily results in lower amplitude and is therefore audible for a reduced distance and is less effective in deterring males and—probably more important—attracting females.[46] (Either of these techniques for lowering their call frequency resembles the situation in which snapping shrimp could, in theory, evolve chelae that are larger than would otherwise be expected for their body size; this does not appear to have happened.)

In canyon wrens, a species of North American bird, territorial males who are exposed to the song of a would-be territorial intruder actually lower the pitch of their own song, apparently to present a more imposing threat to the interloper![47]

What about mammals? Here, deepness of voice isn't obviously constrained by body size. In human beings, at puberty the size of the larynx increases substantially in boys relative to girls, so evidently there is room for human males (or females) to have evolved deeper or higher voices regardless of how large they are.[48] The German opera singer Thomas Quasthoff stands four feet four inches tall, and yet has one of the world's most impressive bass baritone voices. The larynx of male hammerhead bats is three times the size of females, filling nearly all of the males' chest cavity. In some species at least, it is therefore possible for males—regardless of their overall size—to evolve deep voices. Nonetheless, no correlation has been found between larynx size and total body size in British red deer.[49] The roars of these animals—perceived as a threatening challenge by other males—are themselves energetically expensive, which may constitute their own reliability component, thereby serving as a bar to excessive lying.

To some extent, body size can be signaled by the duration of vocal threats, because how long a call can be maintained varies with the size of the lungs. But this, too, can be faked, by the elaboration of air sacs, or—analogous to the trade-off between call frequency and intensity in amphibians—accepting a trade-off between duration of call and its amplitude. As a result, individuals whose calls are more prolonged and thus more intimidating would therefore have to settle for lesser amplitude and therefore reduced carrying distance.[50]

Among primates, the voice can be deepened in a variety of ways. One is to lengthen the vocal tract, either by adding a resonance chamber at the upper end, as is done by proboscis monkeys, whose comically elongated nose (found in males) correlates closely in dimension with body and testes size, and with the size of the harem he is likely to accumulate. Females and juvenile proboscis monkeys have notable noses too, but they are proportionately smaller and pertly upturned, whereas a male's schnoz typically dangles lower than his mouth.

Another voice-lowering tactic is to enlarge or lengthen the vocal tract at the other end by lowering the position of the larynx in the throat, as occurs in human beings. Yet another option, also used by *Homo sapiens*, is to enlarge the larynx itself, as occurs in boys especially, beginning at puberty. As it grows, the front of the larynx is protected by enhanced development of the thyroid cartilage, whose bulk also contributes to the lowered voice tone of men.

This is the Adam's apple, which, as far as we know, is not a Lamarckian inheritance derived from Adam's early and regrettable gustatory habit, after which a piece of the forbidden fruit is alleged to have lodged in his throat. Rather, it—along with a penchant for facial hair—seems likely a result of natural selection enhancing the ability of men to successfully threaten other men. And perhaps, not coincidentally, attract women.

It seems intuitive, at least to human beings, that effective threat vocalizations would be not only low-pitched but also loud. And yet, there are some perplexing exceptions, especially in the avian world. Among song sparrows, for example, the only signal that predicts an attack reliably is when the provoked territory owner sings . . . not loudly, but softly![51] The first report of "soft song" concerned song sparrows, first studied intensively by the suitably named Margaret Nice.[52] Observations of comparably soft songs have been confirmed for a variety of bird species, including "passerine" (perching) birds,[53] as well as other, larger species,[54] and some mammals as well.[55] Something similar is not unknown among human beings, whereby a quietly uttered warning can be more threatening than one that is bellowed. It is nonetheless a mystery why so many species—including our own—often speak softly when carrying a big threat.

One possibility is that by reducing their vocal amplitude, individuals make themselves less evident to possible predators; but, it turns out that soft threats among birds are no more frequent when predators are abundant.[56] Perhaps making understated threats is beneficial in keeping others from knowing what's going on. If your challenge is successful, then it seems likely you would be happy to have it become common knowledge. But if you fail, best to keep it quiet and thus, private: victory has a hundred parents; defeat is an orphan. So, defenders might be expected to use a loud broadcast song, whereas intruders (more liable to lose) should be more circumspect about what they're doing. Alas, this reasonable prediction, as with predator avoidance, is not supported by the evidence.

It is also possible that a loud song gives neighbors a heads-up that a fight is imminent, whereupon they could possibly intrude and try to cadge a copulation while the territorial male is occupied defending his suzerainty. Might this be a good reason, therefore, to use soft song? Perhaps, except that third-party intruders are *more* frequent when the resident gives a soft song.[57] Maybe loud singing is incompatible with conducting an imminent attack, because to sing loudly, birds open their beaks wide and orient their mouth upward, making it difficult to track the movements of an opponent.[58] Yet another possibility is that loud vocalizations use a lot of energy, and that animals—not unlike people—are well advised to conserve their breath if they are about to do something physically demanding, or threatening to do so. In conclusion, the question of why so many threats are soft and quiet is currently unresolved.

Another unresolved question is why some species have more than one threat display. It is possible that a kind of frequency-dependent selection[ix] has operated in such cases. Here's how it might have worked, in conjunction with the elaboration of dishonest bluffing. Imagine a threat behavior that involves a particular vocalization and/or physical posture associated initially with quick resort to a fight. This behavior would select rapidly for a response by its target, often a retreat, given that the display is a good predictor of subsequent assertiveness. At the same time, insofar as it is effective, the display would also exert selection on the senders, who could well be increasingly predisposed to use it even without intent or expectation of escalating to an actual fight— that is, it could devolve into dishonesty in proportion as it is mistaken by the receivers for the real thing.

Moreover, it could also be mimicked, analogous to viceroys mimicking monarchs, by others with neither capability nor intention of following through. As events and generations unfold, such bluffs could eventually be "called" by the target receivers, whereupon this in turn could select for elaboration of a new and different threat, one that—once again—would start out "real" and honest, only to suffer the same eventual fate as the earlier version. Depending on the costs and benefits of each threat display (which would include its likelihood of being tested), the result would ultimately be the existence of multiple threat techniques, coexisting at an equilibrium. For now, however, this is just speculation.

[ix] This phenomenon was described earlier in the context of Hawks and Doves, in which a tactic does well when rare, and less well when abundant.

SEX, SUBTERFUGE, AND SNOOPERS

Attracting females is a widespread goal among male animals and a process in which females are intimately involved, not just passively as recipients of male signals, but also actively as beings with their own agendas and adaptive preferences. The important thing for our purposes is that these mating preferences by females, regardless of the attraction antics by sex-besotted males, can constitute threats to the immediate well-being of males even as they enhance the fitness of choosy females. Back to amphibians. The mating call of male túngara frogs—widely distributed through Central and South America—consists of two identifiable parts: first comes a sonically complex "whine," typically followed by a variable quantity of "chucks." The number of these chuck calls varies with male body size, and females are more likely to mate with males who use many chucks (indicating a larger and thus more desirable sexual partner) as part of their sexual solicitation.[59]

The foregoing seems well suited to honest signaling—albeit in a sexual, rather than threatening, context—but only if "chucking" is energetically expensive, so that larger males would be more able to make these sounds. But as it happens, the energetic cost of túngara frog vocalization comes mostly from the initial whine (which all males make) rather than the subsequent chucks to which female túngara frogs respond favorably.[60] Nor does the chuck or the whine serve as a threat issued to other males. In this species, the threat is borne by each signaling male who, during the course of his romantic serenade, chucks away some of his personal security because he is vulnerable to fringe-lipped bats, unusual predators who respond to túngara frog vocalizations by homing in on the chuck component and are more likely to make a kill in proportion as a calling male avails himself of a lengthy string of those sounds.[61]

In this case, females, by their preference for males whose mating calls involve sexy and dangerous chucks, induce males to expose themselves to greater predation risk than if sexual selection were not operating. The system is accordingly similar to that of sexual selection among birds such as peacocks, described earlier, but with males induced to engage in risky courtship behavior rather than having evolved metabolically expensive anatomical traits such as fancy and conspicuous tail feathers. Elaborate male ornamentation is always accompanied by an eye-catching behavioral repertoire, although the túngara frog is thus far the only species in which courtship behavior alone—without fancy anatomy—has been tied clearly to increased risk of predation.

The strange case of the túngara frog involves three main actors: a female frog, a male frog, and a fringe-lipped bat. In some other species, three-sided interactions occur in the absence of a predator, instead involving a threatener, the threatened, and a third party—a snooper—who is frequently an additional male who listens surreptitiously to, the primary interaction between threatener and threatened. Intended or not, a threat signal often goes beyond the receiver toward whom it is directed and is intercepted by one or more eavesdroppers. This is hardly unexpected, because most threat signals evolved in the context of a social network rather than merely involving the two primary participants. The nature and reactions of audience members is therefore likely to have been anticipated by natural selection, and it is accordingly no surprise that animal eavesdropping is often consequential.

An especially interesting analysis went beyond the Hawk–Dove model described earlier by including a strategy that is appropriately called "Eavesdropper," who challenges the loser of a contest.[62] One might think that eavesdropping would reduce the amount and intensity of escalated conflicts because a third party, by learning who won and who lost, could confront only those individuals revealed to be Doves, thereby avoiding hurtful contests with Hawks. But, assuming that individuals know they are being watched, or listened to, the model suggests the opposite, and for an interesting reason: in the presence of eavesdroppers, both of the original contestants can be expected to escalate more than otherwise to impress the eavesdroppers and reduce the probability that either of the contestants will be challenged in the future.

Researchers assume that eavesdroppers use information gained by paying particular attention to fraught interactions involving the giving and receiving of threats. This makes sense, because it is presumably one of the main reasons individuals in any social species pay special attention to interactions among their peers.

In one study, territory-holding male red-winged blackbirds were presented with stuffed models of males—simulating an attempted territorial incursion—and their responses were noted and characterized. Then the behavior of other, neighboring birds was observed carefully. These neighbors were significantly more likely to trespass onto the territories of "owners" who failed to attack the intruding model. This confirms that (at least among blackbirds) snooping neighbors are a reality, and also suggests that the consequence of threatening an intruder—or failing to do so—includes its impact on attentive neighbors, which in turn has implications for the individuals being intruded upon, likely inspiring them to up their assertiveness.[63]

It is conceivable, however, that a tendency to take advantage of a seemingly wimpy territory owner might not be based on what the neighboring birds witnessed, but rather because the neighbors had personally interacted with the proprietor some time previously. This possibility was tested in a laboratory study of rainbow trout.

Two competitor fish were placed together at one end of an aquarium while a third— the observer—was at the other end, separated from the pair by a clear partition, which allowed the third fish to observe but not interact directly. The two contestants quickly established a winner–loser dominance hierarchy that was reflected in their postures, after which the observer was paired with the winner as well as with a size-matched and equally dominant stranger. The observer settled its aggressive interactions with these two different individuals—either ending up dominant or subordinate—more quickly, and with less aggressive escalation, when paired with a "familiar" individual (the one who had been previously observed) rather than with an unfamiliar individual who it met for the first time.[64]

Fish, not normally considered to possess acute intellects, are thus able to learn the social status of other individuals by watching them, and to adjust their own encounters accordingly. A similar experiment was run with another fish species particularly known for its aggressive inclinations: Siamese fighting fish, aka bettas. Here, the two original contestants were separated by a glass partition because, otherwise, they would likely fight to the death. When their interactions were limited to vision only, they threatened each other vigorously, with the loser assessed as the one who displayed less and adopted a submissive posture. In addition, a third (observer) fish was positioned so that he, too, could interact visually with both the winner and the loser.

Observers took significantly longer to approach those individuals they had seen win, suggesting the observers responded appropriately to the information they had obtained—namely, which of the two was a tougher opponent. On the other hand, their hesitancy to approach winners might have been because winners behaved in a more confident, intimidating, dominant manner when interacting with the observer, rather than because the observer was hesitant because he was deferring to an individual that he had just seen to threaten more successfully. To test this, observers were paired with both winners and losers in contests they had *not* observed; in these cases, observers didn't behave differently toward those fish that had won or lost earlier—so their hesitation in approaching winners in the earlier test wasn't because the winners behaved differently toward them, but because they were observed to have been successful.[65]

In short, threats can be consequential not only insofar as they result in short-term success or failure, but because they can be observed by third parties, which extends the stakes into the future. National leaders are acutely aware that whether their country wins or loses "face," the impact goes beyond any situation that they immediately encounter.

Among free-living animals birds have been the most frequent subjects of research on threat displays because they are relatively easy to observe and, frankly, because their vocal communication lends itself to a different kind of eavesdropping: by inquiring scientists. Among songbirds, who sing when in possession of a territory and whose songs are intended both to attract females and—even more—to threaten and repel other males, territorial challengers also sing. Strangely, perhaps, an aggressive signal among many songbirds that have a large vocal repertoire is "song-type matching," in which an intruder imitates the detailed auditory pattern used by the resident.[66] In such cases, rather than being a form of flattery, imitation is a (presumably sincere) form of challenge, although it is unclear why this should be perceived as threatening to the resident.

More threatening yet and especially aggressive, as well as more understandably challenging—and something that even an observing human can empathize as being downright annoying—is when an intruder adjusts its song to overlap that of a territory owner. This is analogous to overmarking among scent-marking mammals, as well as the widely observed human phenomenon of verbal interruption, notably the irksome reality whereby men interrupt women significantly more often than women interrupt men, speaking over them in a kind of assumed social dominance. (An especially revealing quantitative finding has been that male Supreme Court justices interrupt their female colleagues significantly more often than vice versa.[67])

A "playback" experiment among free-living European nightingales presented territory owners with two loudspeakers, simulating two intruders—one of which was adjusted to overlap part of the song initiated by the recording played through the other speaker. The resident birds responded more strongly to the loudspeaker that was doing the overlapping than to the one being overlapped, indicating they perceived the former as more threatening than the latter.[68] Similar results were found for black-capped chickadees, a common North American bird species.[69] (It isn't known whether female Supreme Court justices perceive their male colleagues as particularly threatening, or just annoying.)

What about the response of females when their mates have experienced an escalated playback threat (i.e., one involving considerable song overlap) compared to those whose male partners experienced a deescalated threat (one not involving overlap)? Adjacent pairs of great tits, a European species related to chickadees, were exposed either to recordings of escalated, overlapping songs or to less threatening, deescalated, nonoverlapping songs, after which the behavior of the resident's female mate was observed closely. Those females whose mate had been exposed to the more threatening, overlapping songs were more liable to desert their mate and visit the adjacent territory of the male who appeared to have been broadcasting that challenge successfully.[70] In short, eavesdropping isn't only used by potential challengers to assess how a resident responds to an outside threat; it can also be used by already-mated females to assess their social partners by how they respond to such threats.

Moreover, a similar experiment run on black-capped chickadees revealed that females who responded in this manner weren't just visiting. They were likely to mate with the neighboring males who appeared to have issued the more threatening challenges, especially when these challenges were not answered effectively by their own mates.[71] The researchers concluded that females "evaluate male–male contests to inform their reproductive choices," a response that can be expected, over time, to select for mated males responding more vigorously to such challenges, but also for challenging males to attempt to be yet more threatening, given the eavesdropping attentiveness and subsequent sexual proclivities of already-mated females.

Sexually motivated intruder males are nonetheless ill-advised to simply maximize their aggressive threats, hoping to enhance their reproductive prospects. Female animals no less than female human beings can be repelled by indications of excessive pushiness and indications of likely "violence-proneness," probably because of their own risk of being injured by an overly aggressive sexual partner. Thus, male Siamese fighting fish use fewer highly aggressive threat displays when interacting with another male in the presence of a female observer, maximizing the conspicuousness of their displays rather than their aggressiveness as such.[72]

Whether fish or fowl, it remains important, however, not to lose, something about which threatener and threatened are doubtless aware. In yet another study involving fighting fish, males who lost a displaying bout while being observed by a female spent significantly less time engaging in sexual display toward the female who had seen them lose.[73] Over many generations, natural selection could very well have informed such individuals that there are other, less informed fish in the sea.

Having considered threat-giving, receiving, and observing in the natural world, as well as honest versus dishonest communication, the next section turns to the most dangerous and threatening animal of all.

NOTES

1. Bradbury, J.W. and Vehrencamp, S.L. 1998. *Principles of Animal Communication*. Sunderland, Mass.: Sinauer.
2. Pope, C.H. 1958. *Snakes Alive and How They Live*. New York, N.Y.: Viking.
3. Glaudas, X. and Winne, C.T. 2007. Do warning displays predict striking behavior in a viperid snake, the cottonmouth (*Agkistrodon piscivorus*)? *Canadian Journal of Zoology*, *85*(4), 574–578.
4. Peter T. Leeson. 2011. *The Invisible Hook: The hidden economics of pirates*. Princeton, NJ: Princeton University Press.
5. Sergio, F., Blas, J., Blanco, G., Tanferna, A., López, L., Lemus, J., and Hiraldo, F. (2011). Raptor Nest Decorations Are a Reliable Threat Against Conspecifics. *Science, 331*(6015), 327–330.
6. Crothers, L.R. and Cummings, M.E. 2015. A multifunctional warning signal behaves as an agonistic status signal in a poison frog. *Behavioral Ecology, 26*(2), 560–568.
7. Fenton, A., Magoolagan, L., Kennedy, Z., and Spencer, K.A. 2011. Parasite-induced warning coloration: a novel form of host manipulation. *Animal Behaviour, 81*(2), 417–422.
8. Lappin, A.K., Brandt, Y., Husak, J.F., Macedonia, J.M., and Kemp, D.J. 2006. Gaping displays reveal and amplify a mechanically based index of weapon performance. *The American Naturalist, 168*(1), 100–113.
9. Clutton-Brock, T.H., Albon, S.D., Gibson, R.M., and Guinness, F.E. 1979. The logical stag: adaptive aspects of fighting in red deer (*Cervus elaphus* L.). *Animal Behaviour 27*, 211–225.
10. Dawkins, R., and Krebs, J. R. 1978. Animal signals: information or manipulation? *Behavioural ecology: An evolutionary approach, 2*, 282–309.

11. Zahavi, A. and Zahavi, A. 1997. *The Handicap Principle: A Missing Piece of Darwin's Puzzle*. London, U.K.: Oxford University Press.

12. Caro, T.M. 1986. The functions of stotting in Thomson's gazelles: some tests of the predictions. *Animal Behaviour, 34*(3), 663–684; Walther, F.R. 1969. Flight behaviour and avoidance of predators in Thomson's gazelle (*Gazella thomsoni*). *Behaviour,* 184–221.

13. FitzGibbon, C.D. and Fanshawe, J.H. 1988. Stotting in Thomson's gazelles: an honest signal of condition. *Behavioral Ecology and Sociobiology, 23*(2), 69–74.

14. Hirth, D.H. and McCullough, D.R. 1977. Evolution of alarm signals in ungulates with special reference to white-tailed deer. *The American Naturalist, 111*(977), 31–42.

15. Bildstein, K.L. 1983. Why white-tailed deer flag their tails. *The American Naturalist, 121*(5), 709–715.

16. Zuberbühler, K., Jenny, D., and Bshary, R. 1999. The predator deterrence function of primate alarm calls. *Ethology, 105*(6), 477–490.

17. Zahavi, A. 1975. Mate selection: a selection for a handicap. *Journal of Theoretical Biology, 53,* 205–214.

18. Emlen, D.J. 1996. Artificial selection on horn length-body size allometry in the horned beetle *Onthophagus acuminatus* (Coleoptera: Scarabaeidae). *Evolution, 50*(3), 1219–1230.

19. Allen, B.J. and Levinton, J.S. 2007. Costs of bearing a sexually selected ornamental weapon in a fiddler crab. *Functional Ecology, 21*(1), 154–161.

20. Moen, R., Pastor, J., and Pastor, A. 1998. A model to predict nutritional requirements for antler growth in moose. *Alces, 34*(1), 59–74.

21. Sharpe, L.L. 2015. Handstand scent marking: height matters to dwarf mongooses. *Animal Behaviour, 105,* 173–179.

22. Porton, I. 1983. Bush dog urine-marking: its role in pair formation and maintenance. *Animal Behaviour, 31,* 1061–1069.

23. White, A.M., Swaisgood, R.R., and Zhang, H. 2002. The highs and lows of chemical communication in giant pandas (*Ailuropoda melanoleuca*): effect of scent deposition height on signal discrimination. *Behavioral Ecology & Sociobiology, 51,* 519–529.

24. https://www.washingtonpost.com/politics/2020/01/20/president-trump-made-16241-false-or-misleading-claims-his-first-three-years/

25. Barash, D.P. 2003. *The Survival Game: How Game Theory Explains the Biology of Cooperation and Competition*. New York, N.Y.: Henry Holt (Times Books).

26. Enquist, M. 1985. Communication during aggressive interactions with particular reference to variation in choice of behaviour. *Animal Behaviour, 33*(4), 1152–1161.

27. Fugle, G.N., Rothstein, S.I., Osenberg, C.W., and McGinley, M.A. 1984. Signals of status in wintering white-crowned sparrows, *Zonotrichia leucophrys gambelii. Animal Behaviour, 32*(1), 86–93.

28. Rohwer, S. 1985. Dyed birds achieve higher social status than controls in Harris' sparrows. *Animal Behaviour, 33*(4), 1325–1331.

29. Rohwer, S. 1977. Status signaling in Harris sparrows: some experiments in deception. *Behaviour,* 107–129.

30. Caldwell, R.L. and Dingle, H. 1975. Ecology and evolution of agonistic behavior in stomatopods. *Naturwissenschaften, 62*(5), 214–222.

31. Knowlton, N. and Keller, B.D. 1982. Symmetric fights as a measure of escalation potential in a symbiotic, territorial snapping shrimp. *Behavioral Ecology and Sociobiology, 10*(4), 289–292.

32. Geist, V. 1971. *Mountain Sheep: A Study in Behavior and Evolution*. Chicago, Ill.: University of Chicago Press.
33. Clutton-Brock, T.H., Guinness, F.E., and Albon, S.D. 1982. *Red Deer: Behavior and Ecology of Two Sexes*. Chicago, Ill.: University of Chicago Press.
34. Dingle, H. and Caldwell, R.L. 1969. The aggressive and territorial behaviour of the mantis shrimp *Gonodactylus bredini* Manning (Crustacea: Stomatopoda). *Behaviour, 33,* 115–136.
35. Dingle, H. 1969. A statistical and information analysis of aggressive communication in the mantis shrimp *Gonodactylus bredini* Manning. *Animal Behaviour, 17,* 561–575.
36. Hughes, M. 1996. Size assessment via a visual signal in snapping shrimp. *Behavioral Ecology and Sociobiology, 38*(1), 51–57.
37. Hughes, M. 2000. Deception with honest signals: signal residuals and signal function in snapping shrimp. *Behavioral Ecology, 11*(6), 614–623.
38. Adams, E.S. and Caldwell, R.L. 1990. Deceptive communication in asymmetric fights of the stomatopod crustacean *Gonodactylus bredini*. *Animal Behaviour, 39*(4), 706–716.
39. Steger, R. and Caldwell, R.L. 1983. Intraspecific deception by bluffing: a defense strategy of newly molted stomatopods (Arthropoda: Crustacea). *Science, 221*(4610), 558–560.
40. Adams and Caldwell, n 35.
41. Caldwell, R.L. 1985. The deceptive use of reputation by stomatopods, pp. 129–146. In R.W. Mitchell and N.S. Thompson, eds. *Deception: Perspectives on Human and Nonhuman Deceit*. Albany, N.Y.: SUNY Press.
42. Ibid.
43. Wilson, R.S., Angilletta, M.J., Jr., James, R.S., Navas, C., and Seebacher, F. 2007. Dishonest signals of strength in male slender crayfish (*Cherax dispar*) during agonistic encounters. *The American Naturalist, 170*(2), 284–291.
44. Davies, N.B. and Halliday, T.R. 1978. Deep croaks and fighting assessment in toads, *Bufo bufo*. *Nature, 274*(5672), 683.
45. Arak, A. 1983. Sexual selection by male–male competition in natterjack toad choruses. *Nature, 306*(5940), 261; Wagner, W.E. 1989. Fighting, assessment, and frequency alteration in Blanchard's cricket frog. *Behavioral Ecology and Sociobiology, 25*(6), 429–436.
46. Martin, W.F. 1971. Mechanics of sound production in toads of the genus *Bufo*: passive elements. *Journal of Experimental Zoology, 176*(3), 273–293.
47. Benedict, L., Rose, A., and Warning, N. 2012. Canyon wrens alter their songs in response to territorial challenges. *Animal Behaviour, 84*(6), 1463–1467.
48. Van Dommelen, W.A. and Moxness, B.H. 1995. Acoustic parameters in speaker height and weight identification: sex-specific behaviour. *Language and Speech, 38*(3), 267–287.
49. Reby, D. and McComb, K. 2003. Anatomical constraints generate honesty: acoustic cues to age and weight in the roars of red deer stags. *Animal Behaviour, 65*(3), 519–530.
50. Fitch, W.T. and Hauser, M.D. 2003. Unpacking "honesty": vertebrate vocal production and the evolution of acoustic signals, A. M. Simmons, A. N. Popper and R.R. Fay, pp. 65–137. In *Acoustic Communication*. New York, N.Y.: Springer.
51. Searcy, W.A., Anderson, R.C., Nowicki, S. 2006. Bird song as a signal of aggressive intent. *Behavioral Ecology and Sociobiology, 60,* 234–241; Searcy, W.A. and Nowicki, S. 2006. Signal interception and the use of soft song in aggressive interactions. *Ethology, 112,* 865–872.
52. Nice, M.M. 1943. Studies in the life history of the song sparrow II: the behavior of the song sparrow and other passerines. *Transactions of the Linnean Society of New York, 6,* 1–328.

53. Dabelsteen, T., McGregor, P.K., Lampe, H.M., Langmore, N.E., and Holland, J. 1998. Quiet song in birds: an overlooked phenomenon. *Bioacoustics*, *9*, 80–105.

54. Reichard, D.G. and Welklin, J.F. 2015. On the existence and potential functions of low-amplitude vocalizations in North American birds. *Auk*, *132*, 156–166.

55. Brady, C.A. 1981. The vocal repertoires of the bush dog (*Speothos venaticus*), crab-eating fox (*Cerdocyon thous*), and maned wolf (*Chrysocyon brachyurus*). *Animal Behaviour*, *29*, 649–669; Gustison, M.L. and Townsend, S.W. 2015. A survey of the context and structure of high- and low-amplitude calls in mammals. *Animal Behaviour*, *105*, 281–288.

56. Anderson, R.C., Searcy, W.A., Peters, S., and Nowicki, S. 2008. Soft song in song sparrows: acoustic structure and implications for signal function. *Ethology*, *114*(7), 662–676.

57. Akçay, Ç., Anderson, R.C., Nowicki, S., Beecher, M.D., and Searcy, W.A. 2015. Quiet threats: soft song as an aggressive signal in birds. *Animal Behaviour*, *105*, 267–274.

58. Akçay, Ç. and Beecher, M.D. 2012. Signaling while fighting: further comments on soft song. *Animal Behaviour*, *83*(2), e1–e3.

59. Rand, A.S. and Ryan, M.J. 1981. The adaptive significance of a complex vocal repertoire in a neotropical frog. *Zeitschrift für Tierpsychologie*, *57*(3–4), 209–214; Ryan, M.J. 1983. Sexual selection and communication in a neotropical frog, *Physalaemus pustulosus*. *Evolution*, *37*(2), 261–272.

60. Ryan, M.J. 1985. Energetic efficiency of vocalization by the frog *Physalaemus pustulosus*. *Journal of Experimental Biology*, *116*(1), 47–52.

61. Ryan, M.J., Tuttle, M.D., and Rand, A.S. 1982. Bat predation and sexual advertisement in a neotropical anuran. *The American Naturalist*, *119*(1), 136–139.

62. Johnstone, R.A. 2001. Eavesdropping and animal conflict. *Proceedings of the National Academy of Sciences*, *98*(16), 9177–9180.

63. Freeman, S. 1987. Male red-winged blackbirds (*Agelaius phoeniceus*) assess the RHP of neighbors by watching contests. *Behavioral Ecology and Sociobiology*, *21*(5), 307–311.

64. Johnsson, J.I. and Åkerman, A. 1998. Watch and learn: preview of the fighting ability of opponents alters contest behaviour in rainbow trout. *Animal Behaviour*, *56*(3), 771–776.

65. Oliveira, R.F., McGregor, P.K., and Latruffe, C. 1998. Know thine enemy: fighting fish gather information from observing conspecific interactions. *Proceedings of the Royal Society B: Biological Sciences*, *265*(1401), 1045.

66. Tom, M.E., Campbell, S.E., and Beecher, M.D. 2013. Song type matching is an honest early threat signal in a hierarchical animal communication system. *Proceedings of the Royal Society B: Biological Sciences*, *280*(1756) 20122517. http://dx.doi.org/10.1098/rspb.2012.2517.; Burt, J.M., Campbell, S.E., and Beecher, M.D. 2001. Song type matching as threat: a test using interactive playback. *Animal Behaviour* *62*(6), 1163–1170.

67. Jacobi, T. and Schweers, D. 2017. Justice, interrupted: the effect of gender, ideology, and seniority at Supreme Court oral arguments. *Virginia Law Review*, *103*, 1379–1487.

68. Naguib, M. and Todt, D. 1997. Effects of dyadic vocal interactions on other conspecific receivers in nightingales. *Animal Behaviour*, *54*(6), 1535–1544.

69. Mennill, D.J. and Ratcliffe, L.M. 2004. Overlapping and matching in the song contests of black-capped chickadees. *Animal Behaviour*, *67*(3), 441–450.

70. Otter, K., McGregor, P.K., Terry, A.M., Burford, F.R., Peake, T.M., and Dabelsteen, T. 1999. Do female great tits (*Parus major*) assess males by eavesdropping? A field study using interactive song playback. *Proceedings of the Royal Society B: Biological Sciences*, *266*(1426), 1305.

71. Mennill, D.J., Boag, P.T., and Ratcliffe, L.M. 2003. The reproductive choices of eavesdropping female black-capped chickadees, *Poecile atricapillus*. *Naturwissenschaften, 90*(12), 577–582.

72. Doutrelant, C., McGregor, P.K., and Oliveira, R.F. 2001. The effect of an audience on intrasexual communication in male Siamese fighting fish, *Betta splendens*. *Behavioral Ecology, 12*(3), 283–286.

73. Herb, B.M., Biron, S.A., and Kidd, M.R. 2003. Courtship by subordinate male Siamese fighting fish, *Betta splendens*: their response to eavesdropping and naïve females. *Behaviour, 140*(1), 71–78.

Section 2

Individuals and Society

CERTAIN HUMAN POSTURES and facial expressions are easy to read and appear to be cross-cultural universals, used automatically and understood unconsciously in nearly all cultures. A clenched fist or jaw, raised weapon, narrowed eyes: you don't have to be a scientist to recognize the message being conveyed, which mirrors the communication of immediate threat found in many animal species. Of special interest to us at present, however, are those threat messages that seem unique to *Homo sapiens*.

Threats and punishments are prominent, for example, in childrearing. Although "Spare the rod, spoil the child" is out of fashion in polite society, it remains a guiding principle for many, and not only those taking their cue from Proverbs 13:24, "He that spareth his rod hateth his son: but he that loveth him chasteneth him betimes" (The Bible, King James Version). Long considered an effective and often necessary means of socializing children, physical chastisement and even the threat thereof are predictors of a wide range of negative developmental outcomes. The extent of agreement in the research literature on this issue is unusual in the social sciences. Threatening and engaging in severe corporal punishment is associated with increased child aggression, antisocial behavior, lower intellectual achievement, poorer quality of parent–child relationships, mental health problems (such as depression), and diminished moral internalization.[1]

There is no question that the cultural river—and not just current research—flows in the direction of "positive childrearing," and thus toward sparing not only the rod, but also threats of the rod. It is nonetheless difficult to avoid the latter, at least when dealing with a recalcitrant child. Few statements are more irritating than "You can't make me," which readily elicits the rejoinder, "Oh yeah?" And only the rare parent has never asserted, "If you do that one more time, I'll . . ." (of course, they rarely actually carry out that threat, which is probably just as well, even though it ensures the child *will* do it one more time). Worse yet is: "If you don't stop crying, I'll give you something to cry about!"

Most expressions of human threat come in the form of "if-then" statements, but not strictly the sort that exasperated parents or professional logicians use. Nor are they based on a simple posture or facial expression, but on language \to communicate the threat: if you steal something, then (in traditional Sharia law) your hand will be cut off. If you steal something in the West, you may well spend time in jail. If you kill someone, you will spend much more time in jail or (in the United States and a handful of other countries) you may be executed. For some religious believers, if you commit an egregious sin, you will be damned forever and spend eternity in hell. This section examines threats and responses to threats as they play themselves out in human interactions— not all such threats because, as mentioned earlier, that task is simply too great. But, none of the topics to follow are insignificant, starting with crimes and punishments.

CRIME AND THE THREAT OF PUNISHMENT

Laws and other rules often deal with bad behavior and what society can do about it using a simple technique: threatening unpleasant consequences if people disobey. Of course, there are many reasons why most people obey most laws. For example, you may simply be safer and better off as a result. Traffic laws often benefit the obedient directly. Drivers in the United States are well-advised to stay to the right, especially on two-lane roads, because other drivers are likely to be doing the same thing and it can be lethally disadvantageous to be disobeying this particular law when encountering a law-abiding oncoming car. Similarly for stopping at a red light, because others, approaching the same intersection at a right angle, will be responding to their green. In such cases, legal threats are generally unnecessary. Obedience is promoted because of the obvious threat that failure to do so can result in an immediately bad outcome.

A second motivation for being law-abiding stems from the widespread feeling that laws and rules should be followed because they carry with them the force of moral obligation. People frequently do things because it is part of being a good and responsible person, although in such cases there is often a behind-the-scenes threat that to behave otherwise is to label one's self deviant, unethical, and possibly at risk of becoming a social outcast or, if nothing else, evoking a painful personal sense of cognitive dissonance.[i] Closely related is what Immanuel Kant labeled the "categorical imperative": the realization by a rational person that one should "always act in accordance with that maxim through which you can at the same time will that it become a universal law." For example, if you are undecided about whether to lie, cheat, or steal, and then ask yourself what society would be like if everyone did that, a rational individual will grasp the answer.

---

[i] A painful psychological phenomenon in which people find themselves confronting two incompatible self-concepts, as with the problem of reconciling belief that you are a good, law-abiding person with the fact that you have just broken a law.

But even here, threats lurk. For one thing, although Kantians prefer to emphasize the strictly rational and supposedly "imperative" aspect of Kant's imperative, it requires not only sufficient cognition to recognize the negative outcome to society if it is not followed, but the negative outcome itself—which good Kantians yearn to avoid—is nothing but a threat; namely, that a community of liars, cheaters, or thieves would become unmanageable and therefore unacceptable. For another, it assumes that people will follow Kant's rule in part because to act counter to one's rational recognition is to threaten one's sense of being rational and considerate.

Finally, laws are also obeyed because (cynics would say nearly *always* because) of the threat of punishment if you get caught. Hence, modern societies that differ in substantial ways are all nonetheless prone to have the equivalent of codified laws and rules, enforced by police, judges, jails, and prisons. These serve many functions: locking up offenders so they cannot repeat their crime, satisfying public desire that wrongdoers be punished, taking this process out of the hands of victimized individuals and their families, and, especially, deterring offenses by the threat of apprehension and incarceration. Threats are thus fundamental to law enforcement, with the expectation that the prospect of "condign" (suitable, appropriate) punishment will not only inhibit malefactors, but also will deter would-be perpetrators by the educative effects of justice administered publicly to convicted offenders.

Whether it does so is another matter. Legally mandated punishments also raise the fraught question of providing equal justice under the law, as well as establishing reasonable laws, suitable punishments, and sometimes—at least—the appearance of mercy. By the end of the eighteenth century, English law was lacking in what we now see as suitable punishments, never mind mercy. It specified 220 different offenses—most of them involving theft of property—that were punishable by death. The expressed intent of the infamous Bloody Code was deterrence: "Men are not hanged for stealing horses," wrote the Marquis of Halifax, "but that horses may not be stolen."[2] But horses were stolen nevertheless, and people—poor people, pretty much exclusively—were hanged for stealing a quill pen or a bolt of cloth.

European reformers, notably the late eighteenth-century Italian jurist Cesare Beccaria, sought a more effective social policy by making, as Gilbert and Sullivan's *Mikado* put it, "the punishment fit the crime." Included in Beccaria's thinking, and unusual in his day, was opposition to the death penalty as not only an improper function of the state, but also ineffective as a deterrent. During the early nineteenth century, English social philosopher Jeremy Bentham, argued strongly for similar adjustments, basically corresponding to "an eye for an eye and a tooth for a tooth," rather than an eye for a tooth or a life for an eye.

It matters greatly whether punishments are just that—a merited consequence of bad actions, regardless of their effect (that is, regardless of the social consequences of those consequences)—or as a deterrent, intended to prevent *others* from performing the same actions in the future. The former, if viewed as an ethical or religious obligation, cannot be evaluated empirically, because their legitimacy is simply mandated from inside (a presumed internal sense of justice that demands to be met) or outside (by religious commandment). On the other hand, the claim that certain punishments and the threat they convey serve as useful deterrents is one that can be evaluated, as we shall see.[ii]

Structured levels of punishment make sense in the world of criminology, not only because it is widely seen as immoral to execute someone for stealing a loaf of bread, but for practical reasons. Before the relaxation of England's Bloody Code, judges and prosecutors often ignored or understated a crime to avoid becoming complicit in overly draconian punishment. Enlightened penology à la Jeremy Bentham and Cesare Beccaria assumed that even potential lawbreakers were "rational actors" who could be deterred by the threat of punishment, as long as the penalty was in line with the crime (hence, credible) and also reliable, prompt, and widely known.

PERMANENT PUNISHMENT

According to Max Weber, one of the founders of sociology, the state is the political entity with an acknowledged monopoly on the "legitimate use of physical force." This includes taking the life of its citizens, presumably according to legalized proceedings and for persons convicted of particularly heinous crimes—notably (although not exclusively), first-degree murder. The death penalty has been justified in different ways, including as a means of protecting society by permanently eliminating especially dangerous transgressors. But it seems likely that the main reason it still exists in much of the United States is not merely that it is seen as the ethically or religiously appropriate response to egregious criminality, but because it satisfies a deep and widespread public demand for retributive justice.

It is probably no coincidence that the United States is the only Western industrialized country in which capital punishment is still practiced, although its popularity is waning. A Gallup survey in 1994 found eighty-percent support of the death penalty, compared to fifty-six-percent support in 2018.[3] Coincidentally or not, the United States is also the most religious Western industrialized country, although religion's popularity in the United States is also diminishing. When Americans were surveyed

---

[ii] The Trump Administration's forcible separation of immigrant children from their parents, when their only "crime" has been attempting to enter the United States *legally*, has been an especially brutal example of a policy officially rationalized for its deterrence value (to prevent other would-be refugees from seeking asylum), but likely motivated by Trump's personal anti-immigrant bias, while also appealing to his base.

about their religious beliefs, starting in the late 1940s, roughly one percent indicated "none." There has been a steady increase in "nones" since then, to about twenty percent by 2018.[4] Within its international peer group, American society continues, nonetheless, to be exceptionally punitive—a likely consequence of a persisting Judeo-Christian orientation that has long emphasized punishment of sin. And nothing is more punitive than executing someone.

Whatever the real reason for the persistence of capital punishment in the United States—and no one really knows—it is most often rationalized not as punishment and definitely not as revenge ("justice" is always the preferred explanation), but as providing deterrence against the worst crimes. The argument is that life-threatening consequences would inhibit potential perpetrators from violating the most basic of society's norms: not to commit treason, rape, and, most conspicuously, intentional homicide.

This claim appears sensible, insofar as reasonable people would seem to avoid actions that could bring about their own death. But, in fact, it is controversial, with the bulk of evidence strongly suggesting that capital punishment does not deter murders. Before examining this hotly debated topic in more detail, let's briefly review one of the most intriguing recent theories: the paradoxical suggestion that socially orchestrated murder—something very much like the modern death penalty—may have acted in our prehistoric past to make us *less* violent than we would otherwise be, at least within our own groups.[5]

People have, historically, been quite fierce—at least on occasion—toward members of a different group. And yet, compared to many other species, we are by and large nonviolent and unaggressive. Thus, nonrelatives and even total strangers typically crowd together on city streets; in a bus, train, airplane, movie theater, or lecture hall; with almost no violence or aggression.

Primatologist Richard Wrangham has proposed that our early ancestors, not unlike many nomadic, nontechnological societies today, were prone to enforcing rules of civil conduct within their group. Unable to call 911 or to employ an independent judiciary and police, they would likely have relied on internal mechanisms for keeping within-group tranquility. In traditional societies today, a trouble-making jerk who consistently disrupts the community typically faces efforts to enforce the accepted social rules by personal warnings, ridicule, or, if necessary, ostracism. But if these gentler attempts are unavailing, and especially if the problem individual is dangerously violent and thus a threat to the group's well-being, capital punishment is frequently agreed upon.

There are few data for how often this form of extrajuridical justice is meted out in contemporary traditional societies, and there is even less evidence speaking to its frequency and impact in human evolution. Nonetheless, the intriguing possibility exists that, as a result of such lethal events, *Homo sapiens* may have actually become less violent, for two reasons. Number one: by removing those especially predisposed to dangerously lethal behavior, the human genome came to harbor fewer of their predisposing genes, making the rest of us less prone to mayhem. Number two: through most of our evolutionary history, groups were small, so individuals knew each other and were also acutely aware of what befell those who conspicuously got out of line. Once such enforcement—lethally administered if need be—became established as cultural tradition, biological selection as well as social pressures and enlightened self-interest would have favored conformity to expected norms. In sum, the idea is that threats to society may have led to informal but effective policing of serious misbehavior, either by eliminating perpetrators lethally or because most group members suppressed whatever inner demons might remain within themselves. Perhaps we therefore owe our comparatively benevolent dispositions to a long history of socially enforced capital punishment.

This hypothesis could be interpreted as an endorsement of the death penalty in modern life, but it needn't be. It is likely that our binocular eyesight and the three-dimensional, stereoscopic vision it affords are attributable to our history as forest-dwelling primates, but that doesn't mean we should start climbing trees and leaping from branch to branch. And the fact that our ancestors may well have scavenged the majority of their animal protein doesn't suggest that we should all begin consuming roadkill. Executing malefactors might, *might*, have made us what we are today, for better and worse. But that doesn't mean that we should keep doing it.

Why not? There are many reasons to conclude that use of the death penalty is an abuse of threat, even aside from its dubious role as a deterrent. Thus, it seems—at best—ethically questionable for society to kill people to demonstrate that people shouldn't kill people. It is especially horrific if an innocent person is executed, and yet no system of justice, however carefully administered, is guaranteed error free. According to the highly reliable Death Penalty Information Center, between 1973 and 2019, 166 occupants of death row in the United States have been exonerated.

Here is just one example, reported by Nicholas Kristof of *The New York Times*. In October 1976, Christopher Williams was sentenced to death for murdering a woman whose house he had allegedly entered, shooting her while she was in bed. This decision was reached despite the fact that forensic evidence showed that the victim had been shot from *outside* the house, through her bedroom window, breaking the glass. Moreover, Mr. Williams had several witnesses prepared to swear that he was attending a birthday party at the time the murder occurred. But, he was an indigent black man, "defended" by an incompetent and apparently indifferent public defender who didn't call a single defense witness, or examine the police record! The Florida Supreme Court eventually overturned the death sentence, leaving Mr. Williams sentenced to life in prison, until in 2019, after having been incarcerated for forty-six years for a crime he did not commit, he was set free.[6]

At least he was not executed. The likelihood is that, among the estimated 15,268 persons who have been put to death in the United States and its preceding jurisdictions between 1608 and 2002,[7] quite a few should have been set free, but weren't.

Racial bias is endemic in the American criminal justice system, not the least when it comes to capital punishment. Of 285 cases in which defendants in Washington state were convicted of first-degree murder between 1981 and 2014, jurors were more likely to impose a death sentence (specifically 4.5 times more likely) when the defendant was black than when the defendant was white.[8] This huge disparity occurred even though government prosecutors were, if anything, somewhat less likely to seek the death penalty against black defendants. Similar research in North Carolina found that convicted murderers were 3.5 times more likely to receive a death sentence if they were black than if they were white.[9] This could result from black Americans being, on average, less able to afford high-quality legal defense (i.e., insofar as capital punishment means that those without capital get the punishment), along with racism—both covert and overt—on the part of jurors and society in general. It also turns out that African Americans whose complexions are darker are more likely to get the death penalty than are those with lighter skin.

Then there is the question of remorse. Once convicted, juries generally take the convict's repentance and self-acknowledged guilt into account when deciding whether to execute prisoners or condemn them to life in prison; those showing remorse are less likely to be executed.[10] This seems reasonable, until you consider that innocent people, convicted of a crime they didn't commit, are unlikely to express a great deal of remorse.

Death penalty prosecutions cost substantially more than life imprisonment,[11] a finding that holds for every state, as does the likelihood that money spent this way could be used more serviceably otherwise. Between 1980 and 2015, to execute 13 people, California spent about $4,000,000,000—taxpayer money that could, alternatively, have been invested in various crime-fighting ways known to be effective, such as hiring roughly 80,000 more police officers, who, according to reasonable estimates, would have prevented approximately 466 murders, along with providing other presumed social benefits.[12] Multiply this by those states in which the death penalty is still carried out, and the result is a substantial chunk of public resources. (In early 2019, California's governor placed a moratorium on executions in that state, which had 740 prisoners on death row at the time.)

Those arguing in favor of executions maintain that the cost of capital cases should not be an argument against it, because the law mandates such expenses as a direct result of insistence by death penalty opponents that legal proceedings in capital cases be so drawn-out. Then those same soft-hearted obstructionists turn around and argue disingenuously that because of these expenses, the death penalty should be abolished! Given the number of death row exonerations, however, combined with the death penalty's irrevocability once carried out, society would seem obliged to bend over backward to make sure those accused, even if convicted, are allowed to pursue all avenues that might lead to acquittal or mitigation.

Time now to visit the key justification of the death penalty: that it is necessary to execute offenders to dissuade others from committing a comparable crime. In this regard, more than forty years after its publication, F.E. Zimring and G.J. Hawkins's *Deterrence: The Legal Threat in Crime Control*[13] remains a fundamental definitional and conceptual text, especially when combined with the updated version of Professor Zimring's magisterial book, *The Contradictions of American Capital Punishment.*[14]

A recent survey sought to evaluate the effectiveness of capital punishment as a deterrent to homicides by asking five hundred chiefs of police throughout the United States to rank those factors that were, in their opinion, the most effective in reducing violent crime. In first place was increasing resources for law enforcement, followed by reducing drug and alcohol abuse. Then, in descending order, came attending to family problems and child abuse, improving mental health treatment options, preparing less crowded court dockets, conducting more effective prosecutions, controlling access to guns, reducing gang violence, and finally—least important and last—was the death penalty.[15] In the same poll, only twenty-four percent of the police chiefs agreed that potential murderers, before committing their crimes, take into account the prospect that they might be executed.

Most murderers either do not expect to be caught or do not carefully weigh the differences between a possible execution and life in prison before they act, making it unlikely that the former might cause them to reconsider whereas the latter would not. In addition, murders are often committed impulsively, during thoughtless moments of passion or anger. As someone who presided over many of his state's executions, former Texas Attorney General Jim Mattox remarked, "It is my own experience that those executed in Texas were not deterred by the existence of the death penalty law."[16] In a sense, this is self-evident: those who committed crimes that led to capital punishment clearly were not deterred by the threat of execution! Moreover, supporters of the death penalty are quick to point out that we don't get to celebrate would-be murders that are deterred by the threat of capital punishment, just as we don't see acts of terrorism that have been prevented by law enforcement. Nonetheless, Mr. Mattox makes sense in that, if the prospect of, say, spending one's life behind bars does not deter such acts, it is highly unlikely that the threat of execution will do so.

Capital punishment nonetheless remains politically popular in much of the United States, in part because it feels good to know that when heinous crimes are committed, the consequence can be satisfyingly severe.[17] Capital punishment may or may not be appropriate for the worst crimes, but that is a moral judgment to which science cannot speak authoritatively. On the other hand, whether it actually deters murder and thus helps keep the law-abiding populace safe is a different matter. The fundamental issue becomes not whether the threat of judicially imposed death deters at all (as it might do, in scattered cases), but whether it is more effective than the threat of protracted penal confinement—that is, whether there is sufficient (or, indeed, any) marginal deterrence resulting from capital punishment that compensates for its many disadvantages.

Scholars of the subject are no less skeptical than police officials about this. Experts from the American Society of Criminology, the Academy of Criminal Justice Sciences, and the Law and Society Association were surveyed. More than eighty percent maintained that the current body of empirical research does not support the claim that capital punishment is justified as a deterrent to murder. A similar proportion—greater than seventy-five percent—rejected the alternative interpretation that perhaps the death penalty's failure to deter murders arises because capital punishment is too rarely carried out to be effective, or that current offenders spend too much time on death row for the threat of execution to influence those who have not yet transgressed.[18]

If the threat of "condign punishment" is an effective deterrent, then one would expect a decline in homicides immediately following executions, especially those that received substantial public attention. Yet, there is a growing body of evidence that executions may actually *increase* the frequency of homicides, perhaps because of the brutalizing effect on the public of their government killing killers—something that could act via a variant of copycat killings, or by contributing to a perception in some psyches that human life isn't all that valuable. In the preface to his play *Man and Superman*, George Bernard Shaw wrote: "It is the deed that teaches, not the name we give it. Murder and capital punishment are not opposites that cancel one another, but similars that breed their kind."[iii]

When a study of homicide rates in California compared data from 1952 to 1967 (during which executions of a condemned prisoner took place every two months on average) with rates from 1968 to 1991 (when no executions were carried out in that same state), the average annual increase in homicide rates turned out to be more than twice as high during "execution years" as during "nonexecution years."[19] It seems distressingly likely that, for some marginal, unstable, and violence-prone individuals, "state-sanctioned executions provide an environment conducive to unsanctioned homicides."[20] In short, it is entirely possible that the more executions, the more people will have to be executed.

It is also possible, on the other hand, that execution years and homicide years corresponded because high levels of homicides generated high levels of executions, rather

---

[iii] Here is another relevant Shaw-ism from the same source: "When a man wants to murder a tiger, he calls it sport. When the tiger wants to murder him, he calls it ferocity. The distinction between Crime and Justice is no greater."

than vice versa, although the delay between the former and the latter is quite lengthy, nearly always by many years and sometimes even decades. Moreover, correlation is not causation, a problem that constantly bedevils efforts to isolate the real-world impact of capital punishment. (After all, this is not an issue that lends itself to experiments.) The best one can do is examine a variety of situations that, ideally, are independent of each other and see whether they converge in telling a similar story. Here, the persuasiveness of the execution–homicide correlation is increased by looking at New York from 1907 to 1963, when that state was carrying out more death penalties than any other. On average, homicides increased during the month after each execution[21]—an immediate short-term effect that decayed over time—and then was renewed after the next execution. This is inconsistent with the view that such a correlation occurred because high frequencies of executions were a *result* of high homicide rates, rather than the latter being stimulated by the former.

Uncertainties remain, however, about whether criminal executions increase or decrease homicide rates. Here are just a few: How do we interpret results when counties within a state vary in whether they pursue death penalties? What if murders are reported inconsistently in different geographic locales, with different policies on capital punishment, and/or during different years? What about local or statewide differences in the effectiveness of defense attorneys or prosecutors? What if murder and execution rates are somehow correlated (either positively or negatively) with some third variable, such as the unemployment rate, changes in social expectations, or whether judges are elected or appointed? Does pressure for instituting and exercising the death penalty increase if the murder rate increases? And if so, how quickly do consequences arise?

If executions have a brutalizing effect on the population at large, and if this effect occurs within the executing jurisdiction, it would operate against a deterrent impact (because more executions would lead to more murders), whereas insofar as such an effect acted outside the executing jurisdiction, it might well not be counted at all. Also, if potential murderers respond to the existence of a death penalty in one jurisdiction by going to another where there is no death penalty, this would exaggerate the seeming deterrent effect, unless murders in both regions were included in any assessment.

Here is yet another complicating factor: there are currently few data on length of prison terms imposed short of capital punishment for various categories of murder. If, as seems likely, jurisdictions with the death penalty—because they are "tough on crime" in general—are also more inclined to impose lengthier prison sentences as well as being less concerned about prisoner well-being, it is at least possible that crime rates (including murder) in such locales would be less than in more lenient jurisdictions, a difference that might then be falsely attributed to the death penalty alone. These and other considerations all apply without getting into the neck-high weeds that obscure the surprisingly complex statistics that apply.

Paradoxically, astrophysicists are one hundred percent confident that gravitational waves were produced 1.3 billion years ago by the collision of two black holes more than a billion light-years away, and yet nothing definite can be said about something so important and nearby as whether capital punishment deters murder right here on Earth.

Opinions, controversies, and difficulties aside, what can be said with any confidence about the actual deterrent value of the death penalty? To what extent does threatening death to murderers make people less liable to commit such crimes? In recent years, more and more states in the United States have been abolishing capital punishment or instituting a moratorium on its use—and at the same time, murder rates have been declining consistently. (Once again, it is possible that this decrease is a result of something else; if so, it still doesn't constitute evidence for the death penalty as a deterrent to homicide.)

Conflicting and confusing claims have long been advanced as to whether the death penalty deters murder—some of which, at least, purported to show that it does. In the mid 1970s, a hugely influential analysis by economist Isaac Ehrlich concluded that each execution saved, on average, eight innocent lives in homicides prevented.[22] This work, although intended to be scientifically rigorous (and considered so when published), has since been discredited substantially, not least because it relied on national trends over time and did not distinguish between states retaining and eliminating the death penalty. Subsequent research has shown, moreover, that changes in murder rates by year have been comparable in states with and without the death penalty.[23]

A 2012 report—by far the most authoritative to date—by the National Research Council, an arm of the National Academy of Sciences, concluded there was no credible evidence that capital punishment deterred homicides.[24] The report therefore urged that consideration of deterrence should be eliminated from the death penalty debate, although since this report the evidence has—if anything—become even stronger that such consideration is indeed appropriate and that it points *against* a deterrent effect. Comparing states that have abolished the death penalty with retainers, homicide rates over successive years go up and down in a wavy pattern, among both groups. The increases and decreases are stunningly similar in both, but the rates in death penalty states remain consistently *higher* across a span of forty years, from 1960 to 2000.[25] (Once again death penalty advocates can argue that this pattern occurred because capital punishment was retained in those states that had higher murder rates in the first place.)

The idea that death penalties actually exert a counterproductive effect by increasing homicide rates would, in conclusion, seem to justify a "Scottish verdict" (not proved). At the same time, it is reasonable that supporters of the death penalty as a socially useful threat have an obligation to demonstrate that it is clearly a deterrent. This assuredly has also not been proved. Indeed, available evidence points in the opposite direction. As a result, the collective case for capital punishment as an effective deterrent threat has —or should have—collapsed. For some more of that evidence, read on.

Murder of police officers is prosecuted with particular vigor, with convicted cop-killers especially likely to be executed. Yet, a thirteen-year study comparing states where the death penalty existed and was readily enforced with those in which it had been abolished concluded that it had no effect on reducing the frequency with which police officers were murdered.[26] If anything, the correlation was reversed: "The three leading states where law enforcement officers were feloniously killed in 1998 were California, the state with the highest death row population; Texas, the state with the most executions since 1976; and Florida, the state that is third highest in executions and in death row population."[27]

Again, we find correlations rather than clear-cut patterns of causation. But, as usual, they are impressive and consistent. A review of the FBI Uniform Crime Reports for each year from 1996 to 2017, for example, shows that states without the death penalty have lower homicide rates than those that retained it, although—once again—causation might conceivably work in the other direction, with states experiencing more frequent homicides being more likely to have instituted and maintained capital punishment as a result.

John J. Donohue, III, is a law professor at Stanford University and a recognized expert on the alleged role of the death penalty as a deterrent to murder. In an article published in 2015, he wrote, "[T]here is not the slightest credible statistical evidence that capital punishment reduces the rate of homicide."[28] Donohue points out, for example, that Canada abolished the death penalty in 1976 and did not restore it, whereas the United States abolished it in 1972 after the Supreme Court (in *Furman v. Georgia*) reduced all pending death sentences to life imprisonment, and then, in *Gregg v. Georgia*, reinstituted it the same year Canada abolished it. Yet the pattern of changes in murder rates was very similar in the two countries. (This is separate from the substantial difference in *absolute* murder rates between the two, with the United States an order of magnitude greater.)

Here is yet another comparative pattern, this one involving two "city-states"—Hong Kong and Singapore—both of which are located in southeast Asia, with roughly the same populations and socioeconomic demographics. A research report aptly titled "Executions, Deterrence and Homicide: A Tale of Two Cities," contrasted the two, noting they experienced very similar murder rates even though they differed vastly in their law enforcement response to murder. Here is part of that article's summary:

Singapore had an execution rate close to 1 per million per year until an explosive twentyfold increase in 1994–95 and 1996–97 to a level that we show was probably the highest in the world. Then over the next 11 years, Singapore executions dropped by about 95%. Hong Kong, by contrast had no executions during the last generation and abolished capital punishment in 1993. Homicide levels and trends are remarkably similar in these two cities over the 35 years after 1973, with neither the surge in Singapore executions nor the more recent steep drop producing any differential impact. By comparing two closely matched places with huge contrasts in actual execution but no differences in homicide trends, we have generated a unique test of the exuberant claims of deterrence that have been produced over the past decade in the US.[29]

Once again, research findings are not comforting for those wishing to argue that the threat of capital punishment deters murder. There is more, particularly from comparing jurisdictions that differ with regard to the death penalty but are essentially similar in other respects—for example, New York and Texas. These two states had comparable murder statistics in 1992, as well as incarceration and execution rates. As crime peaked nationally, the two states differed greatly in their responses. Texas ramped up its executions (especially under then-governor George W. Bush) while also undergoing a massive prison buildup, which resulted in one-hundred thousand more prisoners—an expansion that was considerably greater, in per capita as well as absolute numbers, than occurred in any other state. In contrast, although there were no executions in New York from 1992 to 2003, and an incarceration increase that was considerably less than that of Texas, the decline in New York's homicide rates (62.9%) was significantly steeper than that of Texas (49.6%) during the same years.

According to the annual Uniform Crime Reports, compiled by the FBI, the fluctuating US murder rate peaked in 1980 at 10.2 per 100,000 population, then again in 1991 at 9.8, and in 1993 at 9.5, after which it declined until the year 2000, and has then basically stayed steady ever since, hovering between 4.5 and 5.5 per 100,000. It may be significant that the decrease in the national homicide rate leveled off by 2000, just when the number of persons on death row nationwide had peaked—at more than thirty-five hundred. In short, a very large number of people awaiting execution, which presumably should have increased the salience of the death penalty in the mind of potential murderers, correlated with an *end* to what had been a persistent decline in homicides. If capital punishment were a deterrent to homicide, then one would expect the opposite correlation: many people awaiting execution should cause the murder rate to go down. But instead, the pace at which it had been declining stopped and has remained more or less steady ever since.

Crime in New York rose sharply during the crack epidemic, after which it began dropping around 1992. By 1994, however, the still-high frequency of homicides had a substantial political impact, contributing to the Republican takeover of Congress that year, as well as the defeat of New York's Governor Mario Cuomo (who opposed the death penalty) by George Pataki, who favored it, and who promised to restore it if elected—which he formally did, although no one was executed up to 2004, when that state's supreme court declared it unconstitutional.

Similarly, and arguably more consequential for US politics during the early twenty-first century, the same dynamic contributed to the victory of George W. Bush over then-governor Ann Richards in Texas. Supporters of capital punishment pointed to the succeeding decline in murder rates as confirming their viewpoint, although this reduction had already been well underway, continuing a nationwide trend that—independent of the widespread legalization of capital punishment by then—had begun earlier. Moreover, the rate of that decline did not speed up after these and other "tough-on-crime" election victories.

Then there is the case of Manhattan and the Bronx versus Brooklyn. In 1995, just as New York state was about to reinstitute the death penalty, the district attorneys of Manhattan and the Bronx announced publicly that they would use their prosecutorial discretion and refrain from seeking executions. In contrast, the top Brooklyn prosecutor made it clear he would seek the death penalty for capital crimes. These three New York City boroughs are broadly similar demographically. They also share the same citywide police force and legal structure. Yet, as reported by the New York Law Enforcement Agency Uniform Crime Reports, during the ensuing ten years, the murder rates in the two boroughs where the death penalty was not enforced declined by 64.4% in Manhattan (from 16.3 murders per 100,000 to 5.8) and by 63.9% in the Bronx (from 25.1 per 100,000 to 9.1). During that same interval, murder rates in Brooklyn—where the death penalty was still identifiably enforced—also declined, from 16.6 to 9.0 per 100,000, consistent with the widespread decline in homicides nationwide. But Brooklyn's decrease was 43.3%, significantly less than the reductions in either Manhattan or the Bronx (roughly 64% in each).

It nonetheless seems intuitively obvious that stronger penalties should depress crime, with the strongest penalty inhibiting the most serious ones. It appears, however, that criminals are far more concerned about whether they will be caught than about the consequences if they are. Murders that take place as "crimes of passion" are, by definition, not planned in advance and are therefore not susceptible to deterrence by the threat of capital punishment. On the other hand, a sociopath contemplating homicide—if he thought that he[iv] might well get caught—will in all likelihood be no less deterred by the threat of life in prison without possibility of parole than by potential execution. However, someone who doesn't expect to be caught is not going to be deterred by whatever is threatened.

[iv] Statistically, it usually is a "he," especially when it comes to mass murderers and sociopaths.

It is simply impossible to know how a would-be murderer evaluates the objective risks of being caught—an estimate that even psychologically sophisticated researchers find dauntingly difficult to determine. Moreover, in a large number of cases, murderers expose themselves to so much immediate risk that the chance of possibly being apprehended later and (once again, possibly) subjected to whatever society has in store for them would arguably have very little impact.

Samuel Johnson observed that "when a man knows he is to be hanged in a fortnight, it concentrates his mind wonderfully." Not necessarily. Such "concentration" appears especially unlikely when apprehension is improbable (most murderers are not caught) and, moreover, when execution beckons not in a fortnight, but in the distant future.

Being "smart on crime" ought to trump being "tough on crime"—an argument that is all the more potent as the alleged deterrent justification of capital punishment has been increasingly discredited. In any event, the US national mood has been shifting against capital punishment. Between 1976 and 2019, the United States executed 1,493 prisoners, with the maximum being 98 in 1999, and more recently 23 in 2017 and 25 in 2018.[30] As of 2019, the death penalty has been prohibited in twenty-one states and the District of Columbia, and is still legal in thirty-nine. There is much to be said for taking murderers off the street, much less to be said for satisfying a public baying for retributive blood, and nothing to be said for the use of capital punishment by politicians to enhance their prospects of election—except, perhaps, from the limited perspective of those same politicians.

One of the most conspicuous examples of this occurred in 1992, when candidate Bill Clinton, then-governor of Arkansas, interrupted his presidential campaign personally to attend the execution of convicted murderer Ricky Ray Rector, thereby putting to rest the image that Democrats were "soft on crime"—a claim that had dogged 1988 Democratic presidential candidate Michael Dukakis, who opposed the death penalty.

When he was put to death, Mr. Rector was severely mentally disabled, something that under most circumstances would mitigate against his execution. Yet it is distressingly common for mental incapacity to be ignored when it comes to capital punishment. Thus, according to a 2019 report from the Death Penalty Information Center,

> at least 19 of the 22 prisoners who were executed this year had one or more of the following impairments: significant evidence of mental illness; evidence of brain injury, developmental brain damage, or an IQ in the intellectually disabled range; or chronic serious childhood trauma, neglect, and/or abuse.

The high frequency of serious mental illness among condemned murderers makes a kind of ghastly sense. Just as legal codes generally insist that to be responsible for a crime a perpetrator must be able to distinguish right from wrong, to be deterred by the legal threat of execution, a would-be murderer must be able to recognize the supposed deterrent effect of capital punishment. This recognition is liable to be impaired or altogether absent in someone with severe mental illness.

Even many people who are basically sane and seemingly well-adjusted are not deterred by the threat of execution. "Some are surprised at the thought that the death penalty would not deter murder," writes law professor John Donohue. He goes on:

> They might themselves instinctively feel distinct unease at the thought of doing anything that could lead to a sentence of death. Yet this represents the wrong calculus, since virtually all such individuals would feel tremendous unease at doing something that could put them in a cage for the rest of their lives. The important arithmetic of the death penalty is that it can only have a possible useful effect on a very small number of individuals—those that would not be deterred by the prospect of life without possibility of parole but would be deterred by the presence of the death penalty. In other words, if we look at New York—a state with no capital punishment (as of 2004), a large population (19,300,000) and a relatively low murder rate (4.77 per 100,000 people)—we find that 921 murders occurred in 2006. Assuming that 921 roughly represents the number of murderers in New York in 2006, then this represents the maximum number of individuals whose behavior could have been changed in a socially acceptable manner by the presence of a death penalty law (at least under a rational actor model). But against these 921 murderers who might potentially have been deterred by capital punishment, there were about 19,299,000 individuals in New York who were not deterred by the threat of capital punishment (since it was nonexistent and yet they still did not kill). This number is roughly 20,000 times as great as the number of murderers in New York in 2006. If the death penalty has a brutalization effect, then we at least have to think about whether any of the 19,299,000 current non-murderers might be subject to a malign influence of capital punishment that would work in opposition to any possible benign influence that could potentially influence only 921 individuals.[31]

The United States remains the only Western democratic country to retain capital punishment, whatever its questionable role as a deterrent, and regardless of its possible malign influence. As of late 2018, 170 countries have abolished the death penalty, and 11 more in which it is legal haven't executed anyone for a decade or more. In Europe, only Belarus still executes criminals. According to Amnesty International, by far the number-one executing country is China, with more than one thousand executions in

2017, followed by Iran with more than five hundred, then—in order—Saudi Arabia, Iraq, Pakistan, Egypt, Somalia, the United States, Jordan, and Singapore.[v]

In early 2019, the sultanate of Brunei announced that it would begin enforcing its new penal code, which made adultery and gay sex punishable by death from stoning, plus codifying the amputation of limbs for theft, and public flogging for abortion. (The latter is to be carried out "with moderate force, without lifting his hand over his head . . . and shall not be inflicted on the face, chest, stomach or private parts.") Responding to an international outcry, Brunei's foreign affairs minister wrote to United Nations human rights officials that these and other penalties focused "more on prevention than punishment," going on to explain that the "aim is to educate, deter, rehabilitate and nurture rather than to punish."[32] Having failed as a deterrent for murder in other countries, it remains to be demonstrated how Brunei's threats of capital punishment for either adultery or homosexuality will succeed, or how carrying out said threats will contribute to rehabilitation or nurturing.

In conclusion, the death penalty is an example of attempts to minimize a threat (especially homicide) by use of another threat (executing criminals) that almost certainly has failed in its goal, while also being horribly counterproductive, having resulted in many additional deaths and perhaps even increasing the frequency of heinous crimes. But the death penalty isn't the only example of failed deterrence orchestrated by government at the societal level. Take the peculiar and little-known case of the "chemists' war" in the United States, when the federal government ended up poisoning thousands of Americans in a failed effort to deter them from drinking alcohol.[33]

It began with prohibition, which itself started in 1919, and was, to put it mildly, a failure. Millions of Americans wanted to drink, and managed to do so, illegally, by purchasing "moonshine" either directly from bootleggers or in the tens of thousands of speakeasies that popped up around the country. In addition, enterprising entrepreneurs as well as many thirsty, low-income citizens quickly discovered that inexpensive industrial alcohol—ethanol, used to produce many commercial products, from paints to perfume—could be consumed instead. The feds responded with a legal requirement that alcohol be "denatured" by adding methyl alcohol (methanol), the lethal kissing-cousin of ethyl alcohol (ethanol), the active ingredient in alcoholic beverages.

Methanol causes blindness, psychotic delusions, and, in sufficient quantity, death. Bootleggers responded by distilling the dangerously denatured product, which

---

[v] Here is a complete list of all countries known to have carried out executions between 2013 and 2017: Afghanistan, Bahrain, Bangladesh, Belarus, Botswana, Chad, China, Egypt, Equatorial Guinea, India, Indonesia, Iran, Iraq, Japan, Jordan, Kuwait, Malaysia, Nigeria, North Korea, Oman, Pakistan, Palestinian Territories, Saudi Arabia, Singapore, Somalia, South Sudan, Sudan, Taiwan, Thailand (2018), United Arab Emirates, the United States, Vietnam, and Yemen. (Amnesty International had no reliable data for judicial executions in Libya or Syria.)

removed most of the methanol; the "renatured" drink was only mildly toxic. The government, in turn, mandated ever higher levels of poison; legislation in 1926 doubled the previous level of methanol, so that its toxicity couldn't be distilled away. Also legally required was the addition of benzine and a derivative of kerosene, noxious in taste and odor, and conveying their own health hazards—notably, cancer.

The increased methanol level was widely publicized by the government as a deterrent: at the time, a government chemist explained to *The New York Times* in 1926 that the increase in methanol level was instituted because "It gives a greater warning to the drinker that he is getting hold of something that he should leave alone." It was assumed that this particular kind of deterrence would work because the threat of dying would keep people from boozing. But many did anyhow. An estimated ten thousand Americans died as a result of this failure of deterrence. Something similar happened in the 1970s, when the US government sprayed a different poison, Paraquat, on marijuana fields in Mexico. As with the methanol misadventure of the 1920s, Paraquat doubtless deterred some people from smoking marijuana, but once again, deterrence largely failed and the result was widespread suffering—experienced mostly by poor people who couldn't afford unadulterated liquor or unsprayed marijuana.

In 2019, President Trump announced in New Hampshire "The only way to solve the drug problem is through toughness, and toughness should include executing drug dealers." This claim, in addition to being demonstrably false, illustrates several important aspects of the current drug crisis—and of political psychology. For at least some segments of the population, being tough is equivalent to being manly, which in turn induces some politicians (especially, perhaps, those who feel a special need to prove their manliness) to issue dark and violent threats. Also, socially disruptive behavior that originates in complex social circumstances is widely yet incorrectly thought to be modifiable by announcing such threats.

Not to be forgotten, finally, is the long and bloody cross-cultural history of stupendously gruesome executions carried out in public, ostensibly to deter heinous crimes among the spectators. Interested readers might consult *Seeing Justice Done: The Age of Spectacular Capital Punishment in France*, by historian Paul Friedland; those with particular curiosity (and unusually strong stomachs) can familiarize themselves with the lavish torture and execution visited upon one Robert-François Damiens, who had attempted to assassinate Louis XV. These and similar instances of public executions, officially designed to deter would-be malefactors, turned into occasions of widely popular public entertainment, whose deterrent effect is unknown. In any event, the underlying motivation for such spectacles was probably less deterrence than to demonstrate the power and authority of the state.

Lynchings in the postbellum American South, although officially illegal, were widely condoned by local authorities, largely because they did something similar: demonstrating the power and authority of the white majority to intimidate "uppity" blacks by murdering them, and doing so in public. It is possible that where capital punishment persists in the twenty-first century, the goal of intimidating some or all of the citizenry continues to take precedence over any other function, even though this is never publicly acknowledged.

Of course, when it comes to "reforming" criminals, there also remains the myth that sticks are more effective than carrots, and that mandating compassionate treatment, prevention, and alternative service requirements are for sissies and, moreover, that frightened, frustrated, angry, and despairing people can be threatened into shaping up. Or, as the rueful military announcement goes, "Beatings will continue until morale improves."

RELIGION AS RESPONSE TO THREATS

Religion is no stranger to threats, both as response to them and as a promulgator. In the run-up to the 2008 election, candidate Barack Obama observed "It's not surprising" that when people's lives are difficult and haven't been improving, they "get bitter, [and] cling to guns or religion." This angered many and probably diminished his election victory somewhat, but he was quite right. Later, we consider guns; for now, religion—albeit more briefly than it deserves.

Throughout most of human history (and prehistory), survival was a dicey proposition, and wealth and security even more so. Political scientist Ronald Inglehart has made a convincing case that, under those circumstances, social pressures put a premium on group solidarity, antagonism to outsiders, and adherence to rigid norms and to authoritarian leaders. Especially important for our purposes, religious commitment emerges as a crucial way of countering existential threats. In his book *Cultural Evolution*,[34] Inglehart reviewed data from more than one hundred countries, demonstrating an impressive correlation between physical, economic, and social insecurity on the one hand and—among other things—religious commitment. In short, religions gain adherents by promising the prospect of protection.

The opposite association also exists. With modernization, as people's sense of personal vulnerability has diminished, along with greater individual confidence has come diminished religiosity. In short, as external threats recede, so does religious affiliation, and vice versa. This pattern is detectable in what are the least-religious modern societies—the Scandinavian countries, which have enjoyed wealth, a strong social safety net, a comparative absence of wars, and generally high levels of national happiness. By contrast, although surveys of religious affiliation have shown an increase in self-described "nones," Americans, whose lives have become less economically predictable and socially secure, have remained far more committed to their religious beliefs.

So have many of those who have suffered through personal tragedy (e.g., enduring the death of a child or receiving a terminal diagnosis), war, or a natural disaster, such as a devastating earthquake, tsunami, or epidemic. It will be interesting to see if the coronavirus pandemic exerts a similar effect.

Although not everyone who experiences distress adheres more closely to religious belief (for example, the widespread Jewish question: "Where was God during the Holocaust?"), on balance religious affiliation increases after trauma.[35] It may be that the perception that God permitted suffering—or worse, imposed it—is liable to produce a turning away from faith, whereas the sense that God has provided protection or rescue for those who came through their ordeals relatively unscathed generates greater affiliation. It can be tempting for those who survived a terrible event to attribute their deliverance to divine benevolence, readily overlooking those (presumably no less worthy) who didn't.[vi] In any event, numerous studies have shown that religion tends to be psychologically helpful in adjusting to stressful life events.[36]

As one oft-cited study reported, "individuals with strong religious faith report higher levels of life satisfaction, greater personal happiness, *and fewer negative psychosocial consequences of traumatic life events* [emphasis added]."[37] This research builds on the now-classic findings of pioneer sociologist Emile Durkheim that religiosity correlates with reduced suicide rates, although it does not discriminate between whether less suicidality occurs because people are actually happier or because most religions prohibit suicide, threatening punishment in the afterlife for those who take their own lives.

An analysis of survey data derived from 1,709 individuals who experienced different degrees of violent conflict in three separate societies—Sierra Leone, Tajikistan, and Uganda—found that, in each case, people who experienced greater direct exposure to wars were more likely to become involved in religious activities, either Christian or Muslim, and this effect persisted even several years after the conflict had ceased.[38]

Americans are no different. A detailed study comparing US Armed Forces personnel in combat roles with their noncombatant counterparts came up with comparable results, concluding that although there are indeed some atheists in foxholes: "[C]ombat assignment is associated with a substantial increase in the probability that a serviceman subsequently attends religious services regularly and engages in private prayer."[39] Once again, trauma appears to increase religiosity, although it is also possible that it reflects peer pressure as well as increased supply-side availability of chaplaincy services, and perhaps actual evangelism as well. An open letter to then-Secretary of Defense Robert Gates in 2010, criticized "the widespread practice of battlefield Christian proselytizing," pointing out: "When, on active duty, our service members sought urgently needed

---

[vi] This parallels the cynical observation by "prolife" advocates that everyone who is "prochoice" has already been born!

mental health counseling while on the battlefield and with the gun smoke practically still in their faces, they were instead sent to evangelizing chaplains."[40]

An alternative view comes from a piece in the conservative *National Review*, which argued,

> The more dangerous the mission, the more vital chaplains are to its success. The nearly 1,400 chaplains in the U.S. armed forces . . . must be on-the-spot counselors to men and women living through a kind of trauma that few civilians will ever experience. They prepare soldiers to kill and to die without losing their souls.[41]

Aside from the well-established impact of religion in helping people adapt to serious life challenges, many studies have confirmed that, on balance, religious affiliation makes people happier. In more technical terms, it increases their "subjective well-being," as reported in surveys and also demonstrated in objective measures such as longevity. Typically, there is a flip side to such findings, pointing not only to the medically documented—although ill-understood—role of placebo, but also to the impact of organized religion in generating satisfaction with conditions that should instead be resisted. This perspective, not surprisingly, is mostly characteristic of the political left, consistent with Marx's characterization of religion as "the opiate of the masses" (more accurately translated "the opium of the people"). This oft-repeated observation from his *Contribution to the Critique of Hegel's Philosophy of Right* warrants reading in its fuller context,[vii] where it is somewhat more sympathetic to the value of religion in providing solace in a world of pain and threat than is usually recognized:

> Religious distress is at the same time the expression of real distress and the protest against real distress. Religion is the sigh of the oppressed creature, the heart of a heartless world, just as it is the spirit of a spiritless situation. It is the opium of the people. The abolition of religion as the illusory happiness of the people is required for their real happiness. The demand to give up the illusion about its condition is the demand to give up a condition which needs illusions.

Marx nonetheless critiques organized religion as essentially a nonpharmaceutical opiate crisis, because most religious traditions (and not just the official state-supported varieties) have generally allied themselves with prevailing social, political, and economic policy, whereby "Render unto Caesar what is Caesar's" has often morphed

---

[vii] Appearing in his *Critique of Hegel's Philosophy of Right*.

into using the religious promise of a rewarding afterlife as an anesthetic, inducing the oppressed to accept otherwise unacceptable present-day conditions. Here is the first stanza and chorus of "The Preacher and the Slave," a song from the Industrial Workers of the World's *Little Red Songbook* (1909), written by fabled labor organizer Joe Hill:

> Long-haired preachers come out every night
> Try to tell you what's wrong and what's right
> But when asked how 'bout something to eat
> They will answer with voices so sweet—
> You will eat, bye and bye
> In that glorious land above the sky
> Work and pray, live on hay
> You'll get pie in the sky when you die.[viii]

Today, preachers are more likely to have buzz cuts than long hair, and a so-called prosperity gospel has increasingly replaced pie-in-the-sky promises that the afterlife will provide surcease from sorrow. But as we have seen, on some level, people are still inclined to believe them: distress and the hope of things getting better—whether in this life or the next—increases religious affiliation. It may not be coincidental that distress and the hope of things getting better, albeit specifically in *this* life, have also motivated another quasi-religious affiliation—namely, Marxism. Susan Jacoby, perhaps our most astute observer of American secularism, points out that Marxism has had many of the characteristics of a traditional religion, satisfying many of the same needs. Hence, many people who left the Communist Party ended up converting to Catholicism, which provided the structure that the Party, unlike other secular organizations, offered in the past.[42]

Given the near-universal presumption that God is omnibenevolent, omnipotent, and omni-knowing, efforts to reconcile the existence of evil and unmerited suffering constitute an immense challenge to theology, which has given rise to the intellectual cottage industry known as "theodicy." In practice, however, the hermeneutic gymnastics of theodicy are often pushed aside following most disasters and the threat of more to come. When huge numbers suffer and die in a tsunami, an epidemic, a war, or—with particular historical consequence for philosophy and literature—the Lisbon earthquake[ix] and the like, those preachers and the religious systems they espouse receive

---

[viii] In some versions (as when sung, for example, by my father), the final line of the chorus is followed by a muttered, "That's a lie."

[ix] The Lisbon Earthquake of 1755—and the ensuing tsunami—on the morning of the Feast of All Saints destroyed nearly all of that city and killed an estimated seventy-five-thousand people,

more adherents rather than fewer. This connection is especially potent when climate disasters cause human injuries and deaths compared to purely economic losses. Every increase of one percent in the number of injuries has been accompanied, on average, by an increase in church attendance by nearly four percent[43]

A nasty possibility arises—namely, terrorizing people with the intent of making them more prone to embrace the dictates of doctrine. One of the most cynical depictions of controlling people by such threats was offered by Hermes in Madeline Miller's 2018 novel *Circe*. Here the god elaborates upon his technique: making mortals suffer and then threatening them with yet more misery to coerce the deity's version of good behavior—namely, abundant sacrifices to the residents of Olympus: "Make him shiver, kill his wife, cripple his child, then you will hear from him. He will starve his family for a month to buy you a pure-white yearling calf. . . . In the end, it's best to give him something. Then he will be happy, and you can start over threatening him again."[44]

OH, HELL!

So far, we have briefly looked at religion as it responds to threats; next: its role in generating them. This will not include such obvious cases as religion-inspired wars, witch-burning, the Inquisition, and so on. We look instead at more or less explicit threats (or warnings, depending on one's perspective) of punishment to come, in the "life" to come.

Western holiday music, ostensibly Christian but largely secular, contains this mild remonstrance: "He sees you when you're sleeping. He knows when you're awake. He knows if you've been bad or good, so be good for goodness sake!" Traditionally, the worst that comes from not being good is a lump of coal in one's Christmas stocking—a threat that might have loomed large for many children, but is less than intimidating for most adults. On the other hand, there is a long tradition, especially in Judeo-Christian theology, of anticipating that evil-doers will be strenuously punished. It is entirely possible that this threat has resulted in behavior that has been more prosocial than would otherwise have been the case, leading many to claim that without God and the threats that he imposes, morality would not exist. Certainly, there are numerous cases of divine retribution announced, with apparent satisfaction, in the Hebrew bible. Here is just a sample.

roughly one third its population. It caused additional tremors among Enlightenment thinkers, including Rousseau and Kant, and stimulated Voltaire to write his satiric novel *Candide*, which made fun of Gottfried Leibnitz's contention that "all is for the best in this best of all possible worlds."

For the sin of disobedience, Adam and Eve were expelled from Eden (Genesis 3:14–24). The Noahic Flood was imposed upon a sinful planet (Genesis 6–7); the people of Sodom and Gomorrah were obliterated because of their indiscretions, chiefly sexual (Genesis 19:23–29); ditto for Onan because of, as one might expect, onanism (Genesis 38:6–10); plagues were visited upon the Egyptians because they ignored the entreaties of Moses (Exodus 7–14); different plagues scourged the Israelites for worshipping the Golden Calf (Exodus 32), and Uzzah was struck dead for having touched the Ark of the Covenant (Samuel 6:1–7).

The Christian Bible also contains many statements of God's retributive inclinations, such as, "Whoever believes in the Son has eternal life; whoever does not obey the Son shall not see life, but the wrath of God remains on him" (John 3:36); "For the wrath of God is revealed from heaven against all ungodliness and unrighteousness of men, who by their unrighteousness suppress the truth" (Romans 1:18); and "Let no one deceive you with empty words, for because of these things the wrath of God comes upon the sons of disobedience" (Ephesians 5:6).

The idea that God's wrath can be discerned in natural disasters, from epidemics to floods, hurricanes, and so forth, although denied in some quarters, is nonetheless widespread, especially among fundamentalists and evangelicals. For example, televangelist Pat Robertson claimed that the devastating 2010 earthquake in Haiti may well have been a delayed divine punishment for Haitians having made a "pact with the devil" when they overthrew their French slave-holding overlords two centuries earlier. A week after the terrorist 9/11 attacks, when appearing on *The 700 Club*, a television show hosted by Pat Robertson, the Rev. Jerry Falwell announced:

> The abortionists have got to bear some burden for this because God will not be mocked. And when we destroy 40 million little innocent babies, we make God mad. I really believe that the pagans, and the abortionists, and the feminists, and the gays and the lesbians who are actively trying to make that an alternative lifestyle, the ACLU, People for the American Way, all of them who have tried to secularize America, I point the finger in their face and say, "You helped this happen."[45]

To this, Mr. Robertson added, "I totally concur, and the problem is we have adopted that agenda at the highest levels of our government."

One month into the devastation caused by the coronavirus, in mid-March, 2020, Ralph Drollinger of Capitol Ministries and a senior "faith adviser" to the Trump Administration, announced that America was "experiencing the consequential wrath of God."

These assertions of God's wrath in the here-and-now, however controversial, are mild compared with threats of damnation and eternal torment. It is not uncommon to invoke such outcomes in a nonthreatening mode; one might say, just for the hell of it. But, when intended and taken seriously, as they have been for most of the past two thousand years, notably in the Christian and Islamic worlds, they are serious indeed.

Many, perhaps most, readers of this book are likely to be well educated (it is, after all, a book, and, moreover, published by Oxford University Press). Given that there is a well-established negative correlation between education level and religious commitment,[46] it is probable that a large proportion of readers will doubt the depth and ubiquity of belief in hell. However, a 2011 survey found that 14.7% of Americans were "somewhat sure" that hell existed, whereas 48.4% were "absolutely sure."[47] In 2018, a survey by the Pew Research Center on Religion and Public Life[48] found that although more Americans believe in heaven than in hell, more than eighty percent of "highly religious" and "somewhat religious" Americans maintain that hell is a "real place," whereas fewer than five percent of nonreligious Americans believe similarly. In short, insofar as you, dear reader, may scorn hell—and even those for whom it is real—this is likely to be a consequence of the educational and ideological silos in which so many are isolated. It is also clear that belief in hell has declined in modern times,[49] but that is no reason to discount its prominence in the past as well as in the present.

It can be argued that proclaiming an unpleasant afterlife for sinners is merely intended to encourage pious and socially acceptable behavior in this life—the theological equivalent of offering a gold star to a toddler who pees in the toilet. Leon Bloy, a late nineteenth-century French writer and poet, had referred to the oxymoronic "good news of damnation," under the assumption that only the fear of perpetual hellfire would motivate moral behavior. Such encouragement might seem worthwhile if it results in more prosocial behavior. And, indeed, cross-cultural research has found that belief in what Michael Shermer calls a "cosmic courthouse" waiting to condemn miscreants to divine punishment is correlated with lower national crime rates.[50] Moreover, at least among developing countries, greater belief in hell is associated with more rapid increases in gross domestic product growth.[51]

But there is a line, fine or not, between providing a gentle spur to prosocial behavior and sharpening that spur into a laceration. It is one thing to demand, as in Pink Floyd's *The Wall*, "You can't have your pudding if you don't eat your meat," but what of threatening the same child with a beating? Or brandishing the prospect of unending torture? In short, when does encouragement morph into abuse? Sure enough, research reported under the descriptive title "The Emotional Toll of Hell"[52] found that, despite its positive societal effects, belief in hell is associated with lower happiness and life satisfaction, at both the individual and national levels. Hell makes people more likely to follow prescribed norms, but also more liable to be unhappy. Social psychologist Carol Tavris points out (personal communication) that behaviorists would say the same thing about the use of punishment in getting a child to behave. Children might obey right then, but be sullen and unhappy about it, and more likely to misbehave (i.e., sin) when out of immediate parental oversight.

It is possible, although difficult, to confirm that some believers substitute confidence that a righteous God will punish sinners here on Earth for diminished conviction about hell and its terrors. Nonetheless, religious threats have traditionally focused on the prospect of misery after death rather than during life—an improvement from the threat-giver's perspective over Hermes's technique, because the purported punishment cannot be refuted, although it also cannot be confirmed.

In any event, the dangling of postmortem torments as a way of manipulating the living has evoked criticism from, among others, Voltaire, whose sardonic *Philosophical Dictionary* includes the following reply to someone who had the effrontery to question the existence of hell: "I no more believe in the eternity of hell than yourself; but recollect that it may be no bad thing, perhaps, for your servant, your tailor, and your lawyer to believe in it." The narrator goes on to observe the following:

> To those philosophers who in their writings deny a hell, I will say: "Gentlemen, we do not pass our days with Cicero, Atticus, Marcus Aurelius, Epictetus. . . . In a word, gentlemen, all men are not philosophers. We are obliged to hold intercourse and transact business and mix up in life with knaves possessing little or no reflection, with a vast number of persons addicted to brutality, intoxication, and rapine. You may, if you please, preach to them that the soul of man is mortal." As for myself, I shall be sure to thunder in their ears that if they rob me they will inevitably be damned.

Voltaire's personal feelings, however, were quite clear, as evidenced when he asked, "Were the time ever to arrive in which no citizen of London believed in a hell, what would be adopted? What restraint upon wickedness would exist?" To which he answered, "The feeling of honor, the restraint of the laws, that of the Deity Himself, whose will it is that mankind shall be just, whether there be a hell or not." It is an optimistic expectation, but one that most of the world's religious traditions have not embraced. Even many nonbelievers have touted the benefits of religion as a mechanism of social hygiene, allegedly restraining some of humanity's more unpleasant and antisocial impulses. This is in addition to the older tradition of believers holding to the beneficence of threatening hellfire and brimstone as appropriate to the salvific responsibility of religious leadership, not to manipulate those who would otherwise misbehave but because those who sin and do not repent will literally suffer hellfire and brimstone. Such argumentation is similar to the perspective described earlier that valorizes military chaplaincy as a necessary and appropriate ministry to soldiers.

For the classical Greeks and Romans, hell was neither a place of punishment nor a restraint upon wickedness; it was simply the abode of the dead where all mortals ended up, regardless of their merit or lack thereof. Heroes could visit there under special circumstances, hang out and converse with the "shades" of the deceased, and then return—for example, Odysseus in *The Odyssey* and Aeneas in *The Aeneid*, who met Achilles, Agamemnon, Tiresias, and the like. In most of the world's better known religious traditions, by contrast, fear of death has been augmented by additional fears: that the afterlife has particular horrors in store for those who have transgressed in life.

For the ancient Egyptians, one traveled after death through different regions of the *Duat*, which was more a place of judgment than of punishment, although those who failed the former were subject to the latter—notably, being chomped and then swallowed by *Ammit*, the devourer of souls, who had the rear end of a hippo, the upper body of a lion, and the head and jaws of a crocodile. This unpleasant experience could be avoided, however, by successfully passing the "weighing of the heart," during which Osiris—Lord of the Underworld and Judge of the Dead—assessed each dead person's heart, comparing it with the "feather of *Maat*," a stand-in for fairness and truth. Those whose hearts were heavier than the feather (which presumably was rather light, and thus a challenging threshold) were in big trouble,

In a related telling, one might escape this outcome by uttering a series of denials, such as "I have not cheated," "I have not blasphemed," and so forth. It is unclear how intimidating early Egyptians found either version of the postmortem evaluative process, because there does not appear to be any evidence that anyone ever ended up consigned to Ms. *Ammit*. In the great majority of other religious traditions, some version of the West's conception of hell can be found, complete with explicit threats that associated torture with failure to live up to social rules and expectations. Some versions of Buddhism and Taoism competed by literally multiplying their visions of hell, with each proposing different numbers, in a kind of posthumous poker. For example, the Buddhists promoted eight different levels of hell, to which the Taoists responded with ten, after which the Buddhists countered with eight cold ones and eight hot ones, and the bidding went on, mostly in the double digits, until the Buddhists jumped to eighty-four thousand, whereupon the Taoists gave up.

One version of Burmese Buddhism had (and for some devotees still has) 40,040 different hells, each associated with its own misbehavior, such as one for not returning borrowed books and another for throwing shards of pottery over a wall. It is problematic to generalize about any of the world's great sacred traditions; consider, for example, the differences within Christianity between Roman Catholicism and, say, Unitarianism or Quakerism. But with that caveat, here is a generalization: insofar as most Eastern religions identified a version of hell, they usually mandated a prolonged duration, but short of eternity. One story in Tibetan Buddhism posits a great cube of sesame seeds one hundred miles on a side, from which a small bird removes one seed every thousand years; when the seeds are all gone, the sinner gets a kind of parole to try living their life again.

Better known in the West, of course, is the Buddhist and Hindu warning that misbehavior generates bad karma, which in turn can lead to rebirth as an insect or a snake, which presumably is unpleasant enough, albeit not quite as loathsome and certainly not as lengthy as the eternal damnation with which Christians and Muslims are threatened. Like corporal punishment of misbehaving children, the threat of hell is somewhat out of fashion these days, but it has long enjoyed (if that is the right term) a lengthy history as a fear-inducer, persisting in the hot, painful imagination of millions of psychologically traumatized people.

Many thinkers, preceding Voltaire, disputed the notion that moral behavior depends on the promulgation of threats (hell) versus the promise of rewards (heaven). Aristotle's *Nicomachean Ethics* explored the origin and maintenance of justice, morality, and happiness without once venturing into discussion of heaven or hell. In many African folk traditions, hell has been more a regrettable situation than a place of physical torment, and less a physical place than a state of being ignored by one's descendants.

The Jewish perspective is closer to this conception of hell as a kind of nothingness than as understood in the Christian or Islamic world. It contains many descriptions of God's power, evidently intended to erase any doubts as to the advisability of complying with divine commands, but it has no explicit mention of hell as is found in the other two major monotheistic traditions, although the Hebrew Bible identifies a realm called *sheol*: a circumstance of oblivion and nonspecific darkness. Translated into Greek, it became *hades*, but there are conflicting views among Jewish scholars whether *sheol* should be seen as a place of punishment or merely a situation experienced by all dead souls, regardless of their virtue or iniquity.

In contrast to the views of ancient Egyptians, Eastern traditions, and those of the Hebrew bible, the Christian perception of hell has long been quite explicit and dire. "Do not be afraid of those who kill the body but cannot kill the soul. Rather, be afraid of the One who can destroy both soul and body in hell" (Matthew 10:28). Jesus subsequently goes on to leaven the "good news" of the gospels with dire warnings about a place of darkness where "the worm dies not, the fire is not quenched, and there shall be weeping and gnashing of teeth" (Matthew 9:48).

In the sinister valley of Gehenna, Moloch's followers allegedly sacrificed their victims, a place considered so real that Jesus repeatedly threatened his own followers with punishments that were similar, or worse: "If you say, 'You fool,' you will be liable to the hell [*Gehenna*] of fire," he declared in the Sermon on the Mount (Matthew 5:22). And the synoptic gospels attribute this warning to the Savior many more times:

It is better for you to lose one of your members, than for your whole body to be thrown into hell [*Gehenna*]" (Matthew 5:29). . . . If your eye causes you to stumble, tear it out and throw it away; it is better for you to enter life with one eye than to have two eyes and to be thrown into the hell [*Gehenna*] of fire (Matt. 18:9). . . . If your hand causes you to stumble, cut it off; it is better for you to enter life maimed than to have two hands and to go to hell [*Gehenna*], to the unquenchable fire (Mark 9:43). . . . But I will warn you whom to fear: fear him who, after he has killed, has the authority to cast into hell [*Gehenna*]" (Luke 12:5).[53]

The apocryphal third-century treatise *Apocalypse of Paul* contains much material beloved of hell-mongers. It purports to be an eye-witness account of torments, what Harvard's Stephen Greenblatt calls a "ghastly travelogue" that includes but is not limited to

rivers of fire, insatiable worms, swirling sulfur and pitch, stench, and sharp stones raining like hail on the unprotected bodies of the damned. There are adulterers strung up by their eyebrows and hair; sodomites covered in blood and filth; girls who lost their virginity without their parents' knowledge shackled in flaming chains; women who had abortions impaled on flaming spits. There are virtuous pagans who "gave alms and yet did not recognize the Lord God" and who are therefore blinded and placed forever in a deep pit. Demons—here called the "angels of Tartarus"—carry out special tortures designed for particular types of sinners. Hence, for example, a "lector"—a reader of the lessons in church services—who did not follow God's commandments: "And an angel in charge of his torments arrived with a long flaming knife, with which he sliced the lips of this man and his tongue as well."[54]

The Islamic conception of hell (*Jahannam*), derived from the Qua'ran, contains seven different levels of misery, each successively more intense and intolerable, and each correlated with the severity of the misbehavior being punished. *Jahannam* is well stocked with boiling water, and especially fire (sometimes *Jahannam* is translated as "blazing" or "roaring fire"), along with the tree of *Zagunnum*, which has germinated from the transgressions committed by each malefactor, and is thus tailored individually to each suffering soul. Sinners are forced to consume the fruit of this tree, which in various ways tears at their insides. Islamic scholars are divided over whether the trip to *Jahannam* is one-way or if, eventually—and depending on the misdeeds in question— inhabitants are finally permitted access to heaven.

As mentioned, the threat of hell is somewhat out of fashion these days. But, it should be pointed out—and here is as good a place as any—that this book's discussion of hell and the afterlife is predicated on the assumption that all such notions are pure fantasy. Consistent with scientific open-mindedness, let's posit instead that they are very, very, very unlikely, although because the universe of unlikely events is immense (just short of infinite), it is not unlikely that some events—however unlikely—sometimes occur!

The early church father, Tertullian, who wrote extensively about hell in the second century after Jesus, maintained that, after death, the pious would get to delight in witnessing forever the suffering of the damned—among whom Tertullian included essentially anyone who disagreed with him. A twelfth-century text, the *Vision of Tundale*, enjoyed great currency in its day, not the least (one suspects) because of its explicitly horrifying images of the suffering of the damned—a kind of extreme schadenfreude à la Tertullian. Here is the comeuppance undergone by nuns and especially monks, priests, bishops, archbishops, cardinals, and even popes who violated their vows of celibacy: "The genitals of the men and the women were like serpents, which eagerly mangled the lower parts of their stomachs and pulled out their guts."

Within Christianity, hell had become especially prominent after the fifth century AD, with its horrors becoming more strenuous as the Church faced more dissenters—a situation that began long before the famous protests of Martin Luther. Christian proselytizing usually occurred not by the sword, but by pointing to the gospel (literally, "good news") of Jesus's coming, but also by threatening that nonadherents would, also literally, be "damned to hell." Vigorous debate ensued over whether hell was awaiting those poor souls who had not been saved by Christianity, specifically heathens, who, through no fault of their own, died without having encountered Jesus's teachings. Augustine was particularly unrelenting in this regard, insisting that unbaptized babies, no less than unrepentant sinners, were doomed.

Some support for a more lenient Christian approach to those who predated Jesus comes from the "harrowing of hell," whereby Jesus is said to have descended into hell after his death and resurrection to liberate worthy souls, especially those who had preceded him in life, including—in the case of a famous medieval painting by Benvenuto di Giovanni—Adam and Eve. (The word "harrowing" derives from an Old English word for "plundering," and the main sources for the tale are the Gnostic Gospels and the Apocrypha rather than formal Christian theology.)

In contrast to its eastern, Constantinople-based component, the Roman Catholic Church particularly emphasized that punishment for mortal sins would be fixed and unyielding, with no recourse, perhaps because, given this doctrine, repentance alone was inadequate to avoid hell, which in turn generated a lucrative business via the sale of indulgences. Thus, prior to the Reformation, the Catholic Church literally made money via the threat of hell; indeed, this practice was among those to which Martin

Luther vehemently dissented in *The Ninety-five Theses*, issued in 1517. Then, in 1563, as part of the counterreformation, the Council of Trent decreed that "all evil gains for the obtaining of [indulgences] be wholly abolished," which was formally announced by Pope Pius V in 1567.

A popular focus on hell and its punishments nonetheless continued, of which by far the most renowned and influential was Dante's magnificent poem *The Divine Comedy*, which had been composed two centuries earlier, when the sale of indulgences was in full flower. It is interesting that *The Inferno*, with its exuberantly graphic depiction of the tortures of hell, has always been read more widely and enthusiastically than the other two parts of Dante's masterpiece, *The Purgatorio* and *The Paradiso*, although the latter were written with no less verve and brilliance.

Maybe this is testimony to a deep-seated fascination with the grotesque, combined with a hefty dose of schadenfreude along with concern about what might be awaiting the sinful, even though—in its specificity—*The Inferno* simply reveals the imagination of Dante Alighieri rather than any particular teachings of the Roman Catholic Church. (The following account is probably the result, at least in part, of my own fascination with the imagined grotesque, evidently shared by many readers.)

In addition to the now-familiar torments of burning with fire, being eaten by ravenous beasts, and so forth, *The Inferno* achieves power and even a sort of credibility as an extended warning because of how it often shows punishments fitting crimes—poetically, but also with a gut-wrenching memorability. For example, the suffering of adulterers, notably Paulo and Francesca, is to be blown about by torrential winds that reflect their uncontrolled illicit passion, while simultaneously keeping the transgressors forever apart. Also among the punished, equivocators who refused to take sides in the "Rebellion of the Angels" (derived from the biblical Book of Revelations) are condemned to run about, naked and continually stung by swarms of hornets and wasps, while unavailingly chasing an indistinct banner that presumably represents their constant pursuit, when alive, of their own inconsequential self-interest.

Another cautionary tale presents the comeuppance of fortune-tellers (evidently considered serious malefactors in medieval times), who are forced to walk eternally straight ahead, but with their head on backward. Then there are the politicians who accepted bribes, who find themselves stuck in a lake of boiling hot tar—comeuppance for their sticky fingers—all the while harried by the Malebranche (literally, "evil claws"), which use those claws to rip their flesh if they try to get out of the scalding bath. (Note: The word "Malebranche," although sounding like a singular entity, refers to plural claws, as in "the Malebranche are . . ." .) In addition, we meet a sad collection of hypocrites forced to walk hopelessly along a narrow path, wearing robes that appear lovely to an observer, adorned with shining golden threads—resembling monk habits and thus appearing to be a reward for piousness—but are actually composed inside of unbearably heavy lead. Hence, they manifest hypocrisy made real, painful, and permanent.

The most appallingly suitable tortures are found deeper in *The Inferno*'s bowels. The Sowers of Discord were guilty of ripping asunder that which should have been left intact. Hence, they are dismembered by a ferocious demon, after which their lacerations heal, whereupon they are lacerated once again. Here we also find Mohammed, his body hacked open so that his guts spill out. Interestingly, his son-in-law, Ali, is similarly mutilated because of causing the schism between Shi'a and Sunni. Those generating discord within a family are also divided literally in their own bodies. Consigned to the deepest region of hell (not surprisingly) is Satan, who, in Dante's version, is entombed waist-deep in ice and endowed with three horrible heads, the ones on each side chewing Brutus and Cassius, who are being punished for betraying Julius Caesar, while the middle one eternally masticates Judas, betrayer of Jesus. Satan himself is the ultimate betrayer, and in this particular version of hell, treachery against the established order (whether secular or divine) is the ultimate, irreconcilable sin.

Perhaps the most notable of all poetic justice punishments, however, is observed at a slightly less stygian level, reserved for a pair of former inhabitants of the city-state of Pisa: Count Ugolino and Archbishop Ruggieri. These two had engaged in a sequence of mutual betrayals, and so they are encased in one frozen hole, each gnawing forever on the other's head. Although none of these portrayals represent explicit Church doctrine, they have been immensely influential—not merely as literature, but as the stuff of nightmares.

Those horrors have been somewhat allayed by an intermediate situation that has enjoyed theological traction as a kind of waystation between heaven and hell. According to current Catholic catechism, purgatory is "the state of those who die in God's friendship, assured of their eternal salvation, but who still have need of purification to enter into the happiness of heaven." This ritual purification is widely conceived as somehow involving fire that—whether literally or figuratively—burns away the dross of venial sin. It also satisfies the widespread need of those still alive to do something (typically via prayer) on behalf of their deceased loved ones. After all, there is no reason to pray for those in heaven, and ditto for those consigned irrevocably to hell.

Strange as it seems to secularists—and even to many believers—hell has often been conceived as an actual geographic location, a conception shared historically and even now by some of humanity's greatest minds. As a twenty-four-year-old math genius, Galileo—widely revered as perhaps the preeminent founder of Western science—was approached by the Florentine Academy to calculate some of the quantitative details of hell, using Dante's poem as underlying data. (*The Inferno* enjoyed such a reputation in Galileo's day that many sophisticated scholars took it as literal truth).

In 1588, Galileo presented his results in two lectures to his Florentine sponsors. Using proportional scaling from Dante's poem, Galileo had figured out that Satan was 1,180 meters (3,870 ft) tall and, taking Dante's claim that Satan's naval was at the center of the Earth, he determined the exact depth of hell and the thickness of its dome.[x] Did the great Galileo Galilei believe all this? We'll never know, but it may be significant that what remains of him now is a bony finger—the middle one—from his right hand, on public display in Florence's science museum, where it points upward to the universe or perhaps issues a defiant obscene gesture toward the Church that subsequently persecuted him.

Why do threats of hell enjoy such cross-cultural credibility? After all, they lack a reliability component or anything comparable to the handicap principle that seems so pertinent when it comes to animal threats. On the other hand, they cannot be proved wrong. Unlike the exploits of Greek and Roman heroes, or of the narrator in Dante's vast poem, no one has visited the underworld and returned to tell the tale. By the same token, however, they also cannot be proved correct.

Threats of a punishing afterlife may well satisfy a widespread need to balance the scales of justice, to make the universe fair when our mortal life isn't. Thus, a readily evoked sense of justice demands that the likes of Hitler, Stalin, and Pol Pot get their proper comeuppance, if not in this life, then in whatever comes after.

Or maybe hell-threats piggyback on the credibility of religion in general. It is, in any event, worth repeating that, although belief in hell seems at low ebb in the modern world, the devil and his punishing realm still retain many adherents. Or maybe the persistent credence is a case of what psychiatrist Randy Nesse calls the smoke-alarm principle.[55]

Here it is. We accept the annoying occasional screams of a kitchen smoke alarm when we accidentally burn the toast, because of benefit derived if in fact there is a genuine fire. Analogously, our ancestors were likely predisposed to respond with tense alertness to a whispering in their Pleistocene grassland, even though it might be a small rodent, because it could also be a leopard. Better safe than sorry. Better to believe in the legitimacy of hell's smoke alarm—even though it might be false—than to discover, after death and deprived of any recourse, that your personal house really is aflame.

---

[x] As it happened, he made a mistake in calculating the thickness of hell's dome, because, having determined its span, he simply scaled up his estimates from the most famous dome existing in Florence at the time. But, as Galileo later realized, to maintain adequate strength, the thickness of a supporting structure must increase more rapidly than simple linear extrapolation from its width: the thickness cubed divided by the span squared must stay constant. Galileo's square-cube law is still used by structural engineers today, so regardless of whether he got hell right, he at least obtained a hell of a useful result.

But a metaphor is like a cookie: tasty, but if squeezed too hard, it crumbles.[xi] The fear of eternal damnation isn't a well-intended but hypersensitive smoke alarm or highly attuned awareness of a leopard's exhalations. Although some people may indeed be induced to more prosocial behavior by threats of postmortem misery, it seems likely that many have also been consigned by them to lives of guilt, shame, and anxiety, not to mention sheer terror.

After all, detailed representations of the torments of hell have not been limited to Dante, whose imaginative genius is today read more as brilliant, entertaining literature than as a triptych of serious warnings. Similar threats persist today, notably dilated upon in other works of fiction intended to portray reality:—not so much theological truth as how such threats are conveyed to believers, notably children. One need not accept the formal doctrine of misery following the Last Judgment to fear the outcome, although emphasizing the threat certainly helps. Iconic examples of such threats include what have been called "kinetic sermons," which are designed to terrify their listeners by detailing the eternal horrors awaiting those who depart from official teachings and are intended to get their listeners to fall on their knees and repent, to accept Jesus, to acknowledge their personal iniquity— whatever it takes to obtain for themselves a get-out-of-hell-free card. Here is a sample of the terrifying, gruesome threats recalled by James Joyce, as the young Stephen Dedalus in his autobiographical novel, *A Portrait of the Artist as a Young Man*.[xii] It takes the form of a kinetic sermon.

[. . .] Last and crowning torture of all the tortures of that awful place is the eternity of hell. Eternity! O, dread and dire word. Eternity! What mind of man can understand it? And remember, it is an eternity of pain. Even though the pains of hell were not so terrible as they are, yet they would become infinite, as they are destined to last forever. But while they are everlasting they are at the same time, as you know, intolerably intense, unbearably extensive. To bear even the sting of an insect for all eternity would be a dreadful torment. What must it be, then, to bear the manifold tortures of hell forever? Forever! For all eternity! Not for a year or for an age but forever. . . .

Ever to be in hell, never to be in heaven; ever to be shut off from the presence of God, never to enjoy the beatific vision; ever to be eaten with flames, gnawed

<hr />

[xi] Consider this a metametaphor: a metaphor about a metaphor (and please don't squeeze it too hard, either!).

[xii] In Joyce's telling, young Stephen—who had recently lost his virginity to a prostitute—is utterly mortified, horrified, guilt-ridden, and grief-stricken by this vision of his future comeuppance, which induces him to connect all the more strongly with the Church . . . for a time.

by vermin, goaded with burning spikes, never to be free from those pains; ever to have the conscience upbraid one, the memory enrage, the mind filled with darkness and despair, never to escape; ever to curse and revile the foul demons who gloat fiendishly over the misery of their dupes, never to behold the shining raiment of the blessed spirits; ever to cry out of the abyss of fire to God for an instant, a single instant, of respite from such awful agony, never to receive, even for an instant, God's pardon; ever to suffer, never to enjoy; ever to be damned, never to be saved; ever, never; ever, never. O, what a dreadful punishment! An eternity of endless agony, of endless bodily and spiritual torment, without one ray of hope, without one moment of cessation, of agony limitless in intensity, of torment infinitely varied, of torture that sustains eternally that which it eternally devours, of anguish that everlastingly preys upon the spirit while it racks the flesh, an eternity, every instant of which is itself an eternity of woe.

Those who have not been exposed as children to the repeated prospect of being consigned to eternal agony in hell will find it difficult, perhaps impossible, to grasp the impact of such declarations. But anyone seeking to understand—if only intellectually—how these threats have been used and how they have affected their victims would do well to try. In his classic seventeenth-century treatise *Anatomy of Melancholy*, a remarkably prescient account of what today is labeled depression, Robert Burton noted, "If there be a hell upon earth, it is to be found in a melancholy man's heart." It seems that one way to enhance that melancholy is to bring a hell to Earth by threatening hell to pay.

DEALING WITH DEATH

Whatever one's belief in an afterlife, there is universal agreement about what comes immediately after life—namely, death. Equally, there is no doubt that for nonbelievers as well as the devout, death has a special salience, exerting a powerful hold on nearly everyone, so that for many, it is *the* primary underlying threat, although not always acknowledged. Sometimes, in fact, death may be so frightening that people exert considerable effort not only to keep its reality at bay, but even to keep the thought of it under wraps.

At the same time, the United States may well be unusual worldwide in its denial of death, including the degree to which Americans insulate themselves from its reality. Until the Civil War, families nearly always handled their own funeral arrangements, laying out the deceased in their house's most formal room, often called the parlor. But with thousands of soldiers suddenly dying far from home, embalming gained popularity as a way of preserving the bodies of loved ones until they could be returned to their families. Shortly thereafter, home parlors were replaced by funeral parlors, and, along with the increased sophistication of hospitals, Americans not only died away from home but their remains, too, were also dealt with at a distance.[56]

Sigmund Freud maintained that no one genuinely believes in their own death,[xiii] largely because we cannot imagine ourselves not alive. In his 1915 essay, "Thoughts for the Times on War and Death," he suggested that "whenever we attempt to [imagine our death] we perceive that we are in fact still present as spectators." As a result, "in the unconscious every one of us is convinced of his own immortality," which, far from being regrettable, is actually "the secret of heroism Only by burying this profound threat under the deepest layer of denial are we able to function with psychological health." He argued that, although fear of death underlies more of our behavior than we acknowledge, we almost never fear death itself, because, for one thing, we have never experienced it, and for another, it doesn't even survive in our unconscious. Instead, we fear other threats, such as pain, abandonment, or castration—ignoring, it seems, that although everyone has at some time felt pain, one needn't necessarily have been abandoned or castrated to fear either of these.

For all his insights, some of Freud's ideas were downright bizarre, and some of his ruminations on death are no exception. Especially indefensible and thoroughly inconsistent with everything we know of biology is his insistence on a "death instinct" or *Thanatos*, which is alleged to drive living organisms to become inorganic and that in turn is supposedly "explained" by the fact that life came from nonlife and therefore yearns to return to its prior situation. Compared to such an explanation, reading the entrails of chickens seems downright scientific.

The death of someone else is another story, although perhaps one that touches us more directly than is typically acknowledged.—Except for those traditions in which funerary rituals are required for the dead soul to make it to heaven, funerals, although officially designated as focusing on the deceased, are mostly events that help the living come to terms with their loss. It has also been plausibly suggested that mourning is less concerned with grief over the person who has "passed" than with expressing the mourners' repressed death anxiety—a sensation evoked in full threatening efflorescence by a nearby brush with mortality. Unsurprisingly, *memento mori* are more often painful than pleasant. Add to this the widespread assumption that, after death, people's spirits become frightening, and we get a reminder of the worrisome threat posed by death itself.

Ernest Becker, a philosophically inclined anthropologist much influenced by psychoanalysis, was more persuasive than Freud on the subject of death. He provided the

---

[xiii] My aunt, living in Montreal, once announced that if her husband should die, "I'd take the children and move to New York."

impetus behind social psychologists' burgeoning awareness of death awareness. "The fear of death must be present behind all our normal functioning," wrote Becker in his seminal work, *The Denial of Death*,

> in order for the organism to be armed toward self-preservation. But the fear of death cannot be present constantly in one's mental functioning, else the organism could not function. . . . And so we can understand what seems like an impossible paradox: the ever-present fear of death in the normal biological functioning of our instinct of self-preservation, as well as our utter obliviousness to this fear in our conscious life.

Becker went on to emphasize the peculiar situation of human beings, creatures of great intellect and intense self-awareness, part of whose consciousness includes awareness of their utter materiality, which in turn includes the certainty of their unavoidable oblivion—the fact that

> man is a worm and food for worms. This is the paradox: he is out of nature and hopelessly in it; he is dual, up in the stars and yet housed in a heart-pumping, breath-gasping body that once belonged to a fish and still carries the gill-marks to prove it. His body is a material fleshy casing that is alien to him in many ways—the strangest and most repugnant way being that it aches and bleeds and will decay and die. Man is literally split in two: he has an awareness of his own splendid uniqueness in that he sticks out of nature with a towering majesty, and yet he goes back into the ground a few feet in order to blindly and dumbly rot and disappear forever. It is a terrifying dilemma to be in and to have to live with. . . . The knowledge of death is reflective and conceptual, and animals are spared it. They live and they disappear with the same thoughtlessness: a few minutes of fear, a few seconds of anguish, and it is over. But to live a whole lifetime with the fate of death haunting one's dreams and even the most sun-filled days—that's something else.

A devoted cadre of Beckerians, following his lead, have developed terror management theory (TMT). In a nutshell, the "terror" is of death and the "management" refers to the ways people navigate their psychology in light of that fear. "In every calm and reasonable person," wrote Philip Roth in *The Dying Animal*, "there is hidden a second person scared witless about death." The threat is real, the fear is understandable, the outcome inevitable yet incomprehensible, and the precise time and circumstance unknowable. According to TMT, death awareness is an unavoidable, troublesome byproduct of human intelligence plus self-consciousness.

In "Aubade," perhaps the finest poem of all those written on the subject, Philip Larkin noted that when it comes to death, there is "nothing more terrible, nothing more true." He then goes on:

> This is a special way of being afraid
> No trick dispels. Religion used to try,
> That vast moth-eaten musical brocade
> Created to pretend we never die . . .

Larkin soberly concludes that "Most things may never happen: this one will."

Managing this situation is challenging. According to TMT it takes many forms, including such obvious tactics as seeking immortality, either literally via religious belief or symbolically by producing works of art, music, literature, science, politics, or—in the traditional biological way—by producing sufficient numbers of successful offspring, who, presumably, will keep on doing so.

Other phenomena interpreted by TMT as responses to death awareness include a commitment to national greatness, emphasis on the superiority of human beings over animals, enhanced focus on self-esteem as a way of supporting a sense of value and importance in the face of the anxiety produced by knowledge of one's eventual demise, and even a paradoxical preoccupation with killing as a way of generating the impression of controlling death by wielding it against others.[57] The latter is an especially ungenerous perception, but to be alive when others have perished may convey a sense of power, or even invulnerability. At the same time, it could conceivably do the exact opposite, if being so closely associated with the infliction of death increases awareness of its looming universality.

TMT researchers have generated a wealth of empirical support for their extension of Becker's theory-based insights.[58] Among the most frequent research paradigms has been to increase "death salience"—such as by asking subjects to write about the death of a friend or relative, to witness a film featuring a killing, and the like—after which it is typically found that compared to control subjects who are exposed to mortality-neutral stimuli, those for whom death salience has been increased show greater religious and nationalistic commitment, firmer adherence to charismatic leaders, enhanced interest in phenomena associated with symbolic immortality, and so forth.

In *The Fire Next Time*, James Baldwin wrote:

> Life is tragic simply because the earth turns and the sun inexorably rises and sets, and one day, for each of us, the sun will go down for the last, last time. Perhaps the whole root of our trouble, the human trouble, is that we will sacrifice all the beauty of our lives, will imprison ourselves in totems, taboos, crosses, blood sacrifices, steeples, mosques, races, armies, flags, nations, in

order to deny the fact of death, the only fact we have. It seems to me that one ought to rejoice in the fact of death—ought to decide, indeed, to earn one's death by confronting with passion the conundrum of life. One is responsible for life: It is the small beacon in that terrifying darkness from which we come and to which we shall return.

It is a cliché—or should be—that we are *all* dying, and yet whatever one thinks of the claims of TMT, there is no doubt that most people minimize their conscious awareness of this literally existential threat. (With regard to dying, Groucho Marx quipped, "It's the last thing I intend to do.") Serious avoidance is suggested by the extraordinary number of euphemisms that refer to dying: passed, passed away, gone to her reward, met her maker, slipped off, departed, deceased, demised, breathed their last, no longer with us, lost one's life, bought the farm, crossed the Jordan, along with some cruder ones such as bit the dust, kicked the bucket, gone belly up, or croaked. An especially resonant one—agronomically accurate rather than either theological or threatening—comes from the Laymi people of Bolivia, who, according to Sallie Tisdale in her book *Advice for Future Corpses, and Those Who Love Them*, say that a dead person has "gone to cultivate chili peppers."

This smacks of a tradition, not widely followed among the Big Three Abrahamic religions (Judaism, Christianity, and Islam), that seeks to cultivate—not so much chili peppers—but a wise and peaceful acceptance of death. In his *Plum Village Chanting Book*, Buddhist master Thich Nhat Hanh developed "The Five Remembrances":

"1. I am of the nature to grow old. There is no way to escape growing old.
2. I am of the nature to have ill health. There is no way to escape ill health.
3. I am of the nature to die. There is no way to escape death.
4. All that is dear to me and everyone I love are of the nature to change. There is no way to escape being separated from them.
5. My actions are my only true belongings. I cannot escape the consequences of my actions. My actions are the ground upon which I stand."

A similar focus on the desirability of accepting death as inevitable and with equanimity emerged in ancient Greek and, later, Roman philosophical traditions at about the same time that Buddhism was arising in Asia. Preceding and—to some extent—anticipating the Stoics, it argued, as in Plato's *Phaedo*, that the purpose of philosophy is to prepare one for dying, and that "there is a child within us to whom death is a sort of hobgoblin; him we must persuade not to be afraid when he is alone in the dark." In another of Plato's dialogues, *Apology*, we read that this fear of death is indeed the "pretense of wisdom and not real wisdom, being the appearance of knowing the unknown; since no one knows whether death, which they in their fear apprehend to be the greatest evil, may not be the greatest good."

In Socrates's case—later emulated by Seneca—the preparation seemed to work. Greatest of the Stoic philosophers, Seneca wrote in his treatise "On the Shortness of Life" that "He who fears death will never do anything worthy of a man who is alive." Seneca also maintained that life without the courage to face death is slavery—which sounds fine in theory, although it means that, in practice, nearly everyone is enslaved. (Seneca himself was a notable exception, having calmly killed himself when ordered to do so by his former pupil, the Roman emperor Nero.)

Another well-known Stoic, the Roman Emperor Marcus Aurelius, argued that because at any given moment the past is over and the future nonexistent, death doesn't cost us anything, given that all we have is the present, which is fleeting and instantaneous. If that sounds like sophistry, it pretty much is; in any event, it doesn't seem to have generated much comfort among non-Stoics.

In his *Natural History*, Pliny the Elder (writing about the same time as Seneca) railed against the then-prevalent idea that death isn't final, but somehow leads on to a kind of life redux, albeit one spent in either a better or worse place:

> Plague take it, what is this mad idea that life is renewed by death? What repose are the generations ever to have if the soul retains permanent sensation in the upper world and the ghost in the lower? Assuredly this sweet but credulous fancy ruins nature's chief blessing, death, and doubles the sorrow of one about to die by the thought of sorrow to come hereafter; for if to live is sweet, who can find it sweet to have done living?

Pliny goes on to raise an intriguing point: "How much easier and safer for each to trust in himself, and for us to derive our idea of future tranquility from our experience of it before birth!" This last sentence is sometimes interpreted as a paean to heaven. But read in context, Pliny is actually arguing that we might be tranquil about death because it would leave us no worse off than we were in the immense time that transpired *before* we were born. Given that we don't lament our lack of prebirth "experience," why should we fear an equivalent oblivion after death?

There is a tradition beyond the Stoics that claims to welcome death, and not only in cases of martyrdom or release from intractable pain. Part of being a believing Christian is believing that—assuming you have lived a sufficiently pious life and that God concurs—the "saved" will be going to a "better place" and that, moreover, there is a good chance of being reunited there with those loved ones who have "gone before" (assuming, once again, that they too have been chosen for eternal bliss). It is therefore somewhat paradoxical that even most believers mourn the death of their beloveds, rather than celebrating. On the other hand, just as the Bible speaks of "the peace that passeth understanding," Hamlet—without benefit of biblical inspiration—soliloquizes about a "consummation devoutly to be wished."

The overwhelming majority of people, whether religious or just plain melancholy or something altogether different, are nevertheless alone in the dark and mortally afraid of their mortality. In her book *Natural Causes*, Barbara Ehrenreich has detailed the ludicrous extent to which fear of death has driven even the most seemingly stoic individuals to all sorts of silly extremes in the hope of avoiding or at least delaying it: exercising too much, subjecting themselves to unnecessary body scans that not uncommonly lead to even more unnecessary surgeries, partaking of orthorexia (a condition in which sufferers obsess about eating only foods they consider to be perfectly healthy), and so forth. Ehrenreich subscribes instead to the resigned observation, "Eat sensibly, exercise diligently, die anyway."

Although for most people—consciously or not—awareness of death is pretty much pushed aside as we pursue our quotidian lives, one need not be an avowed Becker booster to recognize the ring of truth in his speculations. Most human beings are appalled by anything that might kill us, natural or not. Rather than concentrating one's mind, thoughts of death far more often terrify it. Even Hamlet, for all his alleged wishing for the consummation of death, acknowledges that

> ay, there's the rub;
> For in that sleep of death what dreams may come
> When we have shuffled off this mortal coil,
> Must give us pause: there's the respect
> That makes calamity of so long life; . . .
> who would fardels[xiv] bear,
> To grunt and sweat under a weary life,
> But that the dread of something after death,
> The undiscover'd country from whose bourn
> No traveler returns, puzzles the will
> And makes us rather bear those ills we have
> Than fly to others that we know not of?

It seems likely that, for believing Christians, the most important "good news" of their religion is that just as Jesus rose from the dead, they too will ultimately achieve the consummation they devoutly yearn for: not death à la Hamlet, but to defeat death such that their souls, if not their bodies, achieve immortality. In his "Holy Sonnet 10," John Donne proclaimed

[xiv] In other words, burdens.

Death, be not proud, though some have called thee
Mighty and dreadful, for thou are not so;
For those whom thou think'st thou dost overthrow
Die not, poor Death, nor yet canst thou kill me.

His poem ends, "And death shall be no more; Death, thou shalt die." It may seem churlish to point this out, but Donne did in fact die, while death is still very much with us. And although the evidence is now overwhelming that many different species of animals (including elephants and chimpanzees) are aware of death when it occurs to others, and even appear to mourn when a relative or sometimes an unrelated group member dies, it does not seem that any animal except ourselves is aware that someday we too will do so. Methinks that those who claim that death is, or should be, no big deal for human beings doth protest too much. Despite all those claims about reducing death's valence, the very fact that the Stoics and others have had to work so hard to overcome fear of it speaks eloquently of how real and threatening is that fear. In her poem "Dirge Without Music," Enda St. Vincent Millay wrote:

Down, down, down into the darkness of the grave
Gently they go,
the beautiful, the tender, the kind;
Quietly they go, the intelligent, the witty, the brave.
I know. But I do not approve. And I am not resigned.

Regardless of whether we are resigned, death, alas, has very sound reasons to be proud, for it is indeed mighty and, for most people, dreadful, making it paradoxical that even those who deny its importance go to great lengths to delay if not prevent it, and even the most religiously devout (especially them) are much concerned to establish and reinforce confidence that eternal life is available, unutterably delightful, and just around that final corner.

GUNNING FOR SOMETHING (IF NOT SOMEONE)

One of the greatest threats that Americans live with comes from guns—and from the guns purchased partly in the hope of alleviating that same threat. When, in his inaugural address, Donald Trump complained about "American carnage," he didn't know how correct he was. Nor, evidently, did he care, because he steadfastly refused to do anything about it.

This genuine carnage isn't a matter of factories and jobs going overseas, but of people dying from gun violence—a lethal brew of suicides, murders, and accidents—that, just since 1970, have produced more American deaths (1.45 million) than all US wars since the country was founded (1.4 million).[59] According to Nicholas Kristof of *The New York Times*, "More Americans die from guns every 10 weeks than died in the entire Afghanistan and Iraq wars combined." During an average day in the United States, one hundred people are killed by guns and another three hundred are injured. American carnage indeed.

A consistent pattern in the realm of threats—from animals to people, society, and international affairs—is that threats and fears are intimately linked. Fearing leads to threatening, and threats, in turn (whether real or imagined), lead to greater fear, resulting in a closely coupled system of reciprocal stimulation. In 1994, the majority of gun owners in the United States reported that their primary reason for toting firearms was recreation: hunting and sport shooting. Just two decades later, in 2016, surveys found a significant change. Fully two thirds of gun owners reported they were motivated by fear of violent crime, even though crime rates fell substantially in the intervening twenty-two years. Roughly seventy percent of gun purchases are now handguns, the only purpose of which isn't to hunt, but to kill people at close range.[60] This is unlikely to be the result of an increase in murderous intent on the part of gun owners, but, rather, increased fear—especially fear of violent crime.

Unscrupulous politicians worldwide are adroit at appealing to fear; it is one of the cornerstones of demagoguery. Between 2004 and 2014, violent crime in the United States declined by a whopping 21 percent (from 463 per 100,000 people, to 365 per 100,000). And yet according to a 2016 poll, sixty-one percent of Americans falsely believed that crime had *increased*, whereas only fifteen percent knew that it had actually *decreased*.[61] This misperception may have been due, at least in part, to candidate Donald Trump, whose presidential campaign included a constant drumbeat that crime (perpetrated especially by immigrants) was a growing danger to "ordinary" Americans.

Epidemiologic data are clear that, rather than reducing whatever violent threats may actually exist, guns in the home significantly increase the risk of firearm-caused death and injury. Adam Hochschild writes,

> If reason played any part in the American love affair with guns, things would have been different a long time ago and we would not have so many mass shootings . . . . Almost everywhere else in the world, if you proposed that virtually any adult not convicted of a felony should be allowed to carry a loaded pistol—openly or concealed—into a bar, a restaurant, or classroom, people would send you off for a psychiatric examination. Yet many states allow this, and in Iowa, a loaded firearm can be carried in public by someone who's completely blind. Suggest, in response to the latest mass shooting, that still more of us should be armed, and people in most other countries would ask what you're smoking . . . .[62]

Nor, when it comes to national legislation, does evidence have any effect. In Massachusetts, which has some of America's most restrictive firearms laws, three people per one-hundred thousand are killed by guns annually, whereas in Alaska, which has some of the weakest, the rate is more than seven times as high. Maybe Alaskans need extra guns to fend off grizzly bears,[xv] but that's certainly not so in Louisiana, another weak-law state, where the murder rate is more than six times as high as in Massachusetts.

All developed nations regulate firearms more stringently than the United States. Compared with the citizens of twenty-two other high-income countries, Americans are ten times more likely to be killed by guns. In the past fifty years alone, more civilians have lost their lives to firearms within the United States than have been killed in uniform in all the wars in American history. And yet, Congress has even prevented the Centers for Disease Control from conducting or sponsoring any studies of gun violence. The threat of knowing the facts—or of crossing the National Rifle Association (NRA) —evidently trumps any threat posed by guns themselves. One reason "gun rights" supporters find it so threatening that their firearms might be taken away—or even subjected to commonsense controls—is that, in a sense, much that they had valued has been lost: bicoastal cities thrive while the "Heartland" suffers from job loss and opiate addiction, among other problems.

New Zealand, a country that in 2019 suffered the death of fifty people praying at two different mosques, has more sheep than people. The United States has more guns than people. A fair percentage of these gun owners are hunters, some are violent criminals, and others simply like the experience of firing a lethal weapon—in most cases, in a firing range or abandoned field. But many (the precise number is unknown) keep guns because of fear, largely generated by the fact that so many *other* people have guns. It is a tragic example of how responding to a perceived threat creates even greater threats—in this case, a country that literally shoots itself in the foot, or worse.

One way to conceptualize the problem is via game theory, specifically the so-called Prisoner's Dilemma, which we examine more closely in the context of international affairs in Section 3. For now, look at it this way. Each person has a choice: carry a gun or don't. The best option, for each—and for society—is *don't*. Not only would everyone save money time and anxiety, but no one gets shot. But here is the dilemma: even if you agree that a gun-free world is better than a gun-toting one, you might worry that someone you meet might not feel similarly and might therefore be carrying a weapon. If so, then you could be at a lethal disadvantage. So, you reason that you need a gun too, if only just to protect yourself.

[xv] This was seriously suggested by Betsy DeVos, Trump administration Secretary of Education, during her confirmation hearings in 2017.

But that's not all. Being a rational person (but *only* rational, not moral), interested in doing the best for yourself, you might also be tempted that if someone you are likely to meet will be unarmed, you could take advantage of that person's restraint by carrying a weapon yourself, so that you could intimidate him or her. And so, whether to defend yourself in the event that others are armed, or to take advantage of them if they aren't, you get a gun. In this model, the reasoning works both ways, and as a result, everyone ends up armed. The dilemma is that, as a result, each of you is worse off than if you had figured out cooperatively some way to refrain from carrying guns in the first place— that is, if society had stepped in and instituted some form of responsible gun control.

It *can* happen. And in New Zealand, it did. Immediately after that 2019 mosque massacre, Prime Minister Jacinda Ardern announced "our gun laws will change," and less than a month later, they did: the Kiwis outlawed all assault weapons, instituting a buy-back program to facilitate compliance. All but one of the country's 120 national legislators voted for these changes, which allowed exceptions for commercial pest control operations and for licensed heirloom collectors (who are required to render their items nonoperational, storing crucial parts in a separate location).

In the United States, by contrast—which has the highest rate of gun deaths of any technologically advanced nation—gun laws have generally been *weakened* even as the country has endured massacres in Charleston, South Carolina (nine people at a bible study group); Las Vegas, Nevada (fifty-eight concert-goers); Newtown, Connecticut (twenty-six first-graders and teachers); Orlando, Florida (forty-nine people in a nightclub); Parkland, Florida (seventeen schoolchildren and faculty); Pittsburg, Pennsylvania (eleven Jewish synagogue attendees); Dayton, Ohio (nine deaths); El Paso, Texas (twenty-two deaths)—the last two on the same day in 2019—and, doubtless, by the time this book is published, more. This refers only to mass murders; the total American carnage on a personal, interpersonal, and within-family scale is substantially higher.

In many states, it is easier to buy a military-style assault weapon than to adopt a kitten from a shelter. In every state, the requirements for getting a driver's license are stricter than for having your own gun. This, despite the fact that, in addition to having one of the highest gun-related murder rates in the world, the data are clear that even modest gun laws are effective in reducing gun deaths. Case in point: Australia experienced a mass murder in 1996 when twenty-six people were killed—apparently at random—in Port Arthur, Tasmania, by a lone gunman using an AR-15 military-style automatic weapon. An enraged populace, backed by police chiefs around the country, encouraged lawmakers to act, which they did.

Within weeks of the tragedy, each of the Australian states and territories had banned assault weapons, including semiautomatics, and the national government banned the importation of these devices and also initiated an aggressive (and generous) buy-back program. According to *Fortune Magazine*, "a land of roughneck pioneers and outback settlers, Australia had never embraced much government regulation and certainly not about their guns. This was a land of almost cartoonish toughness and self-reliance, home of Crocodile Dundee and Australian rules football. Here even the kangaroos box."[63]

The result of Australia's crackdown? During the ensuing twenty-two years, there has not been a single mass shooting. By 2014, which appears to be the most recent year offering such data, that country's murder rate was nearly halved, to 1.0 per 100,000 population, which is less than one fifth that of the United States, and of these, just thirty-two were committed with handguns. In addition, gun suicides in Australia dropped by about eighty percent.[64] Australia's population at the time was twenty-four million; during that same year, Chicago—with a population roughly one sixth that of Australia—registered more than five hundred gun deaths.

Two states within the United States offer an interesting contrast, essentially a case of differing experimental conditions. In 1995, Connecticut firmed up its gun licensing laws, establishing "permit-to-purchase" legislation. During the following decade, firearm homicide rates declined by forty percent. It might be argued that this reduction was a result of diminished lethal violence in the country generally; but, during this same period, nongun homicides in Connecticut remained statistically unchanged.[65] Almost certainly, making it more difficult to obtain a gun made it less likely that people would be killed by a gun.

Missouri is the contrasting example.[66] For decades, the Show-Me state had a permit-to-purchase law, requiring a background check that included in-person appearance at a local sheriff's office. Then, in 2007, this law was repealed.[xvi] Between 1999 and 2006, the gun-related homicide rate in Missouri was already 13.8% higher than the national average; after the repeal, from 2008 to 2014, it skyrocketed to 47% higher. Of course, correlation is not causation, and so we are limited to what might be called "back-strapolation"—reasoning chronologically backward for a conclusion rather than extrapolating forward. But in conjunction with the Australia experience, as well as the contrasting example of Connecticut, a strong case is evident. "Within the United States," writes David Hemenway, Director of the Harvard Injury Control Research Center, "a wide array of empirical evidence indicates that more guns in a community leads to more homicide."[67]

In perverse response to its own lethal experience, rather than tightening its gun laws, the Missouri legislature also adopted an amendment to the state's constitution that declared gun ownership an "inalienable right." In that state, a nineteen-year-old

---

[xvi] Background checks in Missouri are still required for in-store purchases, but not for private transactions, including at the ubiquitous gun shows.

cannot buy liquor, but can legally purchase and carry a concealed handgun. The day following a mass shooting in Odessa, Texas, a new law went into effect in the Lonestar State, loosening controls on whether people could bring guns to school parking lots, houses of worship, and the like. And when the next mass murder occurs, legislators—mostly Republican, but including some Democrats (especially those representing rural constituencies)—will doubtless offer the ritual "thoughts and prayers," but nothing else. This, despite the fact that during 2019 alone, more American children were shot to death in schools than US soldiers died in Afghanistan and Iraq combined.

So, why doesn't the United States have safer gun laws? Politics. But why are US politics this way? History, a perverse threat cycle, and the Second Amendment are much of the story. That controversial amendment reads "A well-regulated Militia, being necessary to the security of a free State, the right of the people to keep and bear Arms, shall not be infringed." Gun advocates emphasize the conclusion that "the right of the people to keep and bear Arms, shall not be infringed," whereas supporters of greater restrictions point out that this "right" is explicitly limited to "a well-regulated Militia," rather than to individuals. Another contributor is the widespread American self-image of independent frontiersmen, ready to defend their homestead against marauders.

But why are politicians so reluctant to stanch what is an ongoing national tragedy, especially given that most Americans are not now threatened by raiding Comanches and ravenous wolves? And, moreover, according to a Pew Research Center Poll released in 2019, sixty percent favor stricter gun laws.[68] Fear doubtless remains a key motivation—fear of the threat emanating from the NRA, which has become the most ferocious and powerful single-issue lobby in the country.

For nearly two centuries after the Bill of Rights was adopted, the Second Amendment was rarely, if ever, seen as relating to individual rights (indeed, it was hardly considered at all), until the Black Panthers. This militant group entered the California statehouse in 1967, armed and insisting they had a Second Amendment right to carry weapons with which to defend themselves against police harassment and shootings of African-Americans. This initiated a remarkable chain of events, as threats ricocheted off each other. The sight of armed African-Americans, insisting on their right to be armed, caused panic among white politicians, which promptly led to legislation, initially in California and quickly approved by then-governor Ronald Reagan, *against* carrying firearms. The NRA, too, took a strong stance on gun control: *in favor of it!*

That threat having been addressed, next came anxiety from "legitimate" gun owners (as distinct from black ones) that they might lose some of *their* rights, which gave rise to an internal, highly focused pro-gun rights rebellion by some members of the NRA, which, for decades previously, had been concerned almost exclusively with target practice, education, and gun safety. The new insurgent NRA leadership followed a very different agenda, leading to intense politicization of the Second Amendment, and giving it a new, individual-focused interpretation. Along with establishing the Second Amendment as a major political battleground, and turning the NRA from a relatively obscure gun safety organization into a pro-gun lobbying powerhouse, these events also inverted "gun rights" from something initiated by the radical left-wing Black Panthers to one overwhelmingly embraced by right-wing politicians and conservative activists—an orientation that continues today.

The NRA's current power derives from a complex stew of perceived threats, including a supposed loss of personal power and autonomy, worry that a dictatorial government (either domestic or foreign) will descend upon a freedom-loving population, and dangers believed to be posed by the modern-day equivalent of angry Native Americans, cattle rustlers, and wild creatures—namely, armed criminals and anyone with a dark complexion Freudians can have a field day with the first of these, the second is so absurd as to be unworthy of refutation, whereas the third has at least some plausibility, evidenced by the widespread slogans that "when guns are outlawed, only outlaws will have guns," and that "the only thing that stops a bad guy with a guy is a good guy with a gun." Although both these maxims have a degree of intuitive appeal, there is no reason to think that either is true, and, in fact, there is considerable evidence to the contrary.

In countries where guns are outlawed (more accurately, where access is more difficult), everyone has fewer guns—outlaws as well as law-abiders. And the proposal that school shootings, for example, would be reduced if teachers were armed has been severely criticized by law enforcement officers and the great majority of teachers, who point out that, almost certainly, this would result in greater carnage, not less, if—as would surely happen—students occasionally gain access to these weapons. In addition, some teachers will doubtless be not only inadequately trained, but themselves lacking in self-control, not to mention that . And gunfights risk endangering yet more lives when bystanders get caught in the cross-fire. Moreover, "good guys" would never be as prepared to deal with the sudden onset of gun violence as perpetrators necessarily would be. Nor would the good guys' weapons be as immediately accessible. (And if they were, then the risk of them falling into the wrong hands would be even greater.)Lacking empirical supporting arguments, it remains unclear what explains America's extraordinary resistance to tightening its gun laws. Most likely it is an array of fears: politicians' fear that supporting gun control would threaten their reelection prospects—a threat largely embodied by the NRA, which is powered in turn by a vocal minority's fear of government, fear of neighbors, fear of strangers, fear of personal weakness and vulnerability, a sense of being threatened by pretty much everything and everyone. As a result, Americans are indeed seriously threatened—not by everything and everyone, but by many of those people with guns (often including themselves) who shouldn't have them in the first place. The comparison with other industrialized countries is striking. In 2013, according to the nonpartisan Global Burden of Disease Study, there were 35.5

gun murders per million population in the United States, compared to 4.9 per million in Canada and fewer than one per million in the United Kingdom.[69] In 2012—a year for which good global data are available—gun homicides per million were Australia, 1.4; New Zealand, 1.6; Germany, 1.9; Austria, 2.2; Denmark, 2.7; Holland, 3.3; Sweden, 4.1; Finland, 4.5; Ireland, 4.8; Canada, 5.1; Luxembourg, 6.2; Belgium, 6.8; Switzerland, 7.7; *and the United States, 29.7.* Correlated with this, guns are hugely more abundant in the United States than elsewhere: with 4.43% of the world's population, we have 42% of the world's civilian-owned guns. The rate of gun deaths in the United States exceeds even that of Iraq and Afghanistan.

As for mass shootings among countries with a population greater than ten million, only Yemen exceeds the United States in frequency per capita.[70] Moreover, looking across 171 different countries, frequency of mass shootings varies directly with the proportion of gun ownership in each country. This association is not a result of different countries simply being more violent, because when the data are controlled for differences in homicide rates, the correlation still exists. Moreover, the United States is pretty much average when it comes to crime compared to the other twenty or so high-income countries—but not when it comes to gun deaths. In other words, it isn't that Americans are more violent or murderous than other people; rather, it's that when there's an opportunity for violence, guns are much more at hand.

Nor does the US murder rate appear to be the result of violent video games, contrary to what opponents of gun control like to assert. Extensive research data are inconclusive, but if video games have any effect, it is almost certainly a minor one. Japanese and South Koreans, for example, spend more time and money on such games per capita than do Americans, and yet murder rates in those two countries are dramatically lower than in the United States. (Japan and South Korea, unlike the United States, have stringent gun control legislation.)

China, where civilian gun ownership is strongly prohibited, provides an interesting contrast to the US. Between 2010 and 2012, China suffered roughly a dozen apparently random attacks on schoolchildren. Twenty-five were killed, and in nearly all cases the assailant used a knife, never a gun. During this same time period, there were five mass shootings in the United States, resulting in the death of seventy-eight persons. Given that the population of China is roughly four times that of the United States, its lethality rate, scaled up to the population of China, would equate to 312, which is to say that attacks in the United States were 12 times more deadly. Why? Guns are more deadly than knives.

Or, compare Japan and the United States. During 2013, guns were involved in the following number of deaths in the United States: 505 accidental shootings, 11,208 murders, and 21,175 suicides, for a total of 32,888 gun-related deaths. In Japan, during that same year, guns were the lethal instruments thirteen times. Scaling up the Japanese population to that of the United States (Japan has about one third the US population), the Japanese experience equates to thirty-nine total deaths. Hence, the average American is 843 times more liable to die by gun accidents, murders, and suicides than the average Japanese (32,888 divided by 39)— Japan, not coincidentally, has very strict gun laws and even police are very rarely armed. As ever, of course, correlation is not causation.

Granted, there are some countries—notably impoverished ones with very high crime rates, such as Honduras, El Salvador, and Brazil—with gun homicide rates higher than that of the US, but there is no question that among its "peer"-developed countries, the United States is far in the lead when it comes to both guns per capita and gun deaths. Not only that, but it bears repeating that the US does not have an especially high crime rate—excepting gun-related murders.[71] The reality is that when violent crimes occur in the United States, they are more lethal, because guns are more liable to be used. Comparing a resident of New York with someone in London, each is equally likely to be robbed, but when this happens, fifty-four New Yorkers are liable to be killed for every Londoner.

Another troublesome fact is that some of the killing—especially within the United States—is done by police, ostensibly in performance of their law enforcement duties. In addition to the evident racial bias in such killings,[72] there is this discomfiting prospect: in a society drenched in guns, police are—not surprisingly—more worried about their own safety and, thus, more likely to shoot first and ask later whether their suspect was lethally armed. There is a suggestive correlation between the number of guns in a country and the number of police killings that occur there.

Here are data for 2016 to 2017, with the first number in each case indicating the abundance of guns per 100,000 people and the second, in parentheses, showing the number of police killings: Iceland, 32 (0); Switzerland, 27 (0); Finland, 32 (3); Sweden, 23 (6); Canada, 35 (36); and, for the United States, 121 (and a whopping 996 police killings). In 2018, 1,165 Americans were killed by police; it is only common sense that some proportion of these deaths were precipitated by anxiety on the part of law enforcement officers that they may have been dealing with a gun-toting opponent— anxiety that is almost certainly magnified by the abundance of guns in civilian hands, and this applies regardless of whether the worry is well-founded in any specific case.

The gun drama is the clearest example we have encountered so far of a dangerous positive feedback or vicious circle, when a seemingly appropriate response to a threat turns around and bites the responders, increasing the threat it is expected to resolve. (The most worrisome case, nuclear deterrence, is visited at length in this book's final section.) There appear to be several ways a "gun rights" position can become lethally counterproductive. Without purchase restrictions, more people have guns, which increases the pressure to have them, in part because of social expectations and their function as status symbols. More important, increasing the number of gun owners—at least some of whom are irresponsible—increases the danger to themselves and others, which in turn generates its own vicious circle whereby demand for yet more guns is increased, a kind of Chinese finger puzzle writ large.

Even more directly, having guns in the house—once again, often as a hedge against threats outside—actually increases threats inside: more death and injury from those guns. According to the Harvard School of Public Health, even after controlling for robbery rates, there is a positive association between homicides resulting from firearms and the percentage of a state's population living in a household containing guns.[73] Rather than providing protection against "bad guys," having a gun, in short, is a major risk factor for being killed by a gun—something that can happen in three ways: by homicide, accident, or suicide.

The Transportation Security Administration reported that, in 2018, 4,239 firearms were discovered in carry-on luggage. Of these, eighty-six percent were loaded and fully one third had a bullet in the firing chamber.[74] Although the likelihood is that very few of these involved people intending mayhem, at minimum the numbers reflect extraordinary carelessness on the part of gun owners—serious accidents waiting to happen. And happen they certainly do, mostly to children. A total of 2,549 children younger than the age of nineteen died by gunshot during 2014 alone (a typical year), whereas an additional 13,576 were wounded.[75]

Next: suicide. The sixteen-state National Violent Death Reporting System found that in 2009—again, a typical year—51.8% of suicide deaths in the United States were the result of firearms (as was 66.5% of homicide mortality). The Israeli Defense Forces instituted a policy prohibiting soldiers from bringing their weapons home on weekends; it resulted in a forty percent drop in suicides, notably by younger family members.[76] Suicide rates during weekdays were unaffected, thereby providing a "control group" for an impressive natural, national experiment.

There is a belief that women in particular could benefit by having access to a gun, compensating for their presumed physical limitations and lesser aggressiveness, but a multinational survey found that countries and households with higher female gun ownership had *higher* female gun-related victimization.[77] Moreover, a clear pattern has emerged in the United States. Research titled "Weapons in the Lives of Battered Women," based on interviews with 417 women who obtained refuge in sixty-seven different battered women's shelters, found that when a gun had been present in their household, it was used against the victim 71.4% of the time, in most cases as a threat.[78]

It also turns out that a gun in the home makes it far more likely that an abused woman will be killed by her intimate partner.[79] (It is at least possible that some at-risk women might benefit by a gun in the home, if it were their own, and if their partner didn't have access to it; but, there are no data available to test this supposition.)

The danger of guns in the home is emphasized by the fact that people are far more threatened by one-on-one homicides than mass shootings, which receive the bulk of public attention, although they comprise only a tiny minority of gun-related deaths. For example, there were 78 mass shootings in the United States (defined as 4 or more deaths) between 1983 and 2012, totaling 547 deaths, whereas, in contrast, 11,622 people died from gun homicides during the year 2012 alone.[80]

Another belief, or rather, a myth promoted by the NRA and other gun advocates, is that guns somehow aren't implicated in gun violence: "Guns don't kill people; people do." Those people killers are purported to be mentally ill, whereupon the problem becomes one of psychology or psychiatry, thereby letting the weapons themselves— and their purveyors and political enablers—off the hook. This claim is typically trotted out (along with the ubiquitous "hopes and prayers" for the victims and their families) after attention-grabbing mass shootings, which, as just noted, account for only a miniscule fraction of total gun deaths. But Americans, who suffer a startlingly high proportion of gun deaths, do not have a higher rate of mental illness than occurs in other developed countries. Putting it another way, other developed countries have roughly the same proportion of mentally ill people as the United States, but they don't experience anything like the US rate of gun violence.

In the United States, only about three percent to five percent of violent acts have been attributed to diagnosable mental illness, and most of these do not involve guns.[81] It is simply bogus to blame mental illness for the sky-high rate of American gun deaths. Rather, mass murderers in particular are overwhelmingly angry, hate-filled, and, compared to comparable people in other countries, very well armed. At this point, you pretty much have to be crazy to claim that America's gun violence problem is the result of mental illness.

There is no connection between national suicide rates (one measure of mental illness in a society) and the rate of mass shootings. If anything, the association is the other way: countries with high suicide rates tend to have a *lower* rate of mass shootings.[82] Even mass shootings are not especially weighted toward the seriously mentally ill. A carefully maintained long-term database of 235 mass murderers was reported in 2015 to include 52 persons who could be identified as psychotic. Researcher Michael H. Stone, a psychiatrist at Columbia College of Physicians and Surgeons, concluded that "most mass murders are planned well in advance of the outburst, usually as acts of revenge or retribution for perceived slights and wrongs" and are not associated with diagnosable psychosis.[83] Alcohol is far more to blame,[84] especially when combined with domestic violence and ready access to murderous weapons.

Earlier we briefly considered Barack Obama's comment about the people left behind who cling to religion; his actual statement was about "guns and religion," an observation that applies especially to white America. Had the reference been to religion alone, it could have applied equally to blacks, but with the addition of guns, white America was particularly implicated. There is indeed a racial slant (actually, several) to the firearm bloodbath in the United States, rarely mentioned but statistically real—namely, that black Americans die from guns at more than twice the rate of whites.[85] On the other hand, whites are more than five times more likely than blacks to commit suicide with a gun, whereas for every black who commits suicide with a gun, five are killed by a gun wielded by someone else. Whites are far more likely to have guns; blacks, to die by them, a result of homicide. Why do whites keep so many guns? Nostalgia? A cultural signifier? Bear in mind that the primary issue here is not .30-06 hunting rifles, but revolvers and military-style assault weapons.

But what sort of legitimate hunter would use an AR-15 automatic assault rifle to shoot a deer? Or a rabbit? Or to shoot six-year-olds in a first-grade class? Are guns the white person's security blanket? Is it perhaps racial fear, white anxiety of being threatened by blacks? Racial guilt over a long history of slavery plus years of differential access to the American Dream combined with a kind of subconscious fear of black retribution—a Nat Turner's rebellion for the twenty-first century? A legally confirmed private right to bear arms in the home didn't exist until 2008 when the Supreme Court case of *District of Columbia v. Heller* was decided five to four (and written by Anthony Scalia), after which, as already described, racial anxiety suddenly traced this "right" back to 1776 or thereabouts. Since then, Second Amendment rights have trumped children's rights not to be shot.[86]

Many people thought that the horror of nearly two dozen young children murdered in Sandy Hook, Connecticut, in 2012 would mark a turning point in the United States; it may have been just that, but not as gun control advocates had hoped. Rather, when this extreme provocation did not cause a change in national gun laws, it showed that, in the United States, murdering children was preferable to instituting even such common sense controls as universal background checks, keeping firearms from people known to be violent, outlawing military-style assault weapons and high-capacity magazines, and the like.

Alas, within the United States, there are gun nuts so devoted that at least some believe in conspiracy theories regarding mass murders such as Sandy Hook, some of them claiming that it didn't really happen but is merely a lie perpetrated by those looking for an excuse to confiscate America's guns, and others maintaining that such tragedies are actually false flag operations whereby gun control advocates did the killing to provide a pretext to—once again—deprive law-abiding Americans of their firearms.

As of 2019, assault weapons such as AR-15s and AK-47s were legal in forty-three states, whereas high-capacity magazines were legal in forty-one. Despite regular calls for restrictive, commonsense initiatives after each mass murder outrage, as of 2020 the federal government has not passed any gun control legislation in a quarter century, following a ban on the sale of assault weapons in 1994, which the George W. Bush Administration allowed to expire in 2004.

Once again, evidence for an effect is correlational, but nonetheless highly suggestive. During the decade of the assault weapon sales ban, the number of mass gun killings dropped by thirty-seven percent, in contrast with the ten preceding years when the ban was not in effect. Then, in the next ten years after the ban had expired, mass gun murders skyrocketed, with an increase of 183%, resulting in an upsurge in deaths by an enormous 239%.[87] Assault weapons are designed specifically to fire many bullets in a short period of time. Not surprisingly, their use in mass shootings results in mass casualties. In a 2019 report, they were used in "at least 11 of the 15 gun massacres since 2014; at least 234 of the 271 people who died in gun massacres since 2014 were killed by weapons prohibited under the federal assault weapons ban," but were once more legal after 2004.[88]

Thus far, data of this sort have been ineffective in moving Congress to consider reimposing the assault weapons ban. In fact, the only federal legislation concerning gun violence in the past 25 years made things worse: a 2005 law indemnified gun manufacturers against legal liability for homicides or accidental deaths in which their products are used. Interestingly, makers of children's pajamas can be sued if their product is found to be flammable and associated with burns or death; guns, by contrast, are the only consumer item that when used as intended, kill people.

Here is a relatively new example. Shotguns have long been used by hunters, but they're also very lethal against people. The K57 12-gauge shotgun made by KelTec, isn't for hunting—at least, not for hunting game. Its "bullpup" design is unusual, placing the handle and grip in the middle of the barrel, allowing the accuracy of a long muzzle with the maneuverability of something more like a large pistol. FBI data show that from 1998 to 2017, crime in the United States has declined substantially, with aggravated assault down thirty-one percent, burglary down fifty percent, and larceny down thirty-eight percent, but gun sales have skyrocketed. KelTec's ad for the K57 reads "Mi casa NO ES su casa," and you can bet that this perversion of a well-known, famously welcoming Spanish phrase was not directed at Spanish speakers—at least, not as potential buyers.

To summarize: the United States experiences substantially more gun deaths per capita than any other developed country. This unenviable record is not readily attributable to a generally higher crime rate nor to a higher rate of mental illness, violent video games, or to any identifiable cause other than the fact that the United States is awash in guns. This abundance and the resulting "American carnage" appears to result from a confluence of threats—both perceived (by citizens) and, to some extent, real (by politicians, who feel threatened if they stand against the NRA and some voter preferences). In the process, the gun culture in the United States has generated, counterproductively, an especially gruesome array of its own threats. Insofar as the US gun culture is driven, at least in part, by ethnic, racial, cultural, and tribal insecurity, this suggests our next topic: the threat-driven politics of right-wing populism.

NATIONAL POPULISM AND VICE VERSA

There were two particularly intense geopolitical threats to Western democracies during the twentieth century: Nazism (in Germany) along with fascism (in Italy and Japan), both of which were pretty much eliminated by the Second World War—and communism, which largely self-destructed during 1989 through 1991. This appeared to usher in what was sometimes seen as "the end of history" or, at least, the triumph of liberal democracy.

But during the second decade of the twenty-first century, a third great threat emerged: national populism, and not just in the person of Donald Trump in the United States. Viktor Orbàn in Hungary, Recep Tayyip Erdogan in Turkey, Rodrigo Duterte in the Philippines, Vladimir Putin in Russia, Narendra Modhi in India, and Jair Bolsonaro in Brazil became some of the national exemplars of an international phenomenon that has had manifestations in other countries as well, such as the Brexit campaign in the United Kingdom, and strongly right-wing populist movements in Austria, France, Germany, Holland, Italy, Poland, and elsewhere.

In France, for example, the rise of proto-Nazi populism has been abetted by fear of *le grand remplacement*—anxiety that the country is becoming Islamic—not only because of immigration, but also because of falling birth rates among the "traditional" French population.

A comparable sense of living under threat exists in the United States, reflected in a book by one Madison Grant that appeared in 1916. Titled *The Passing of the Great Race,* it is little known today, but was immensely influential in its time, and was endorsed by Theodore Roosevelt, among others. It warned about "racial suicide" resulting from immigration. In the fevered imagination of that time, Italians, Jews, and Eastern Europeans were seen as among those racially inferior groups threatening "white" suzerainty in the US.(Early in the nineteenth century, immigrant threats were seen to emanate from Irish and Germans—and Catholics in general—whereas later in that century, a modest Chinese influx was perceived as especially threatening.

By the twentieth century, Grant's book was taken up enthusiastically by Adolf Hitler in Germany. It also stimulated the US Immigration Act of 1924, which established immigration quotas from different countries and closed the door on people with Asian lineage. Although eventually rewritten in the 1950s, this law was also endorsed several decades later by then-Senator Jeff Sessions, who was to become, for a time, Donald Trump's Attorney General and one of the most fervent supporters of race-based and ethnically based immigration restrictions.

Similar sentiments powered the notorious white supremacists' "Unite the Right" rally and riot in Charlottesville, Virginia, in 2017, which included the chant "You will not replace us," which quickly morphed into "Jews will not replace us." Immediately afterward, President Donald Trump described this event—in which a counterprotester was murdered and featured threatening neo-Nazi thuggery—as including "very fine people on both sides."

Liberal democracy had promised and provided protection against an array of threats: to personal liberty, freedom of religion, private property, and the like. But in the minds of some of its beleaguered citizens, it failed in other respects—notably when it comes to countering threats to family, local groups, religious communities, ethnicity, and nation—all previously reliable sources of identity that have begun to seem endangered. In treating people equally—or at least, claiming to do so—liberal democracy has also undercut the need of many to feel special in their religion, family, race, ethnicity, family background, social status, and the like. Abortion rights threaten the religious beliefs of those who believe that personhood begins at conception, just as gay rights—notably marriage—threaten those who maintain the traditional perspective that marriage can only be between a man and a woman ("Adam and Eve, not Adam and Steve").

The result has been a turn toward the security thought to be provided by strong leaders and their authoritarian governments. This goes a long way toward explaining, for example, why so many soybean farmers continued to support President Donald Trump even when his trade wars with China were so economically hurtful to them. Such persistence is puzzling to economists and other social scientists who assume that political affiliations derive from a rational assessment of self-interest—the old trope of *Homo economicus*—but it makes sense when we consider that *Homo sapiens* aren't always so sapient.

A dominance-oriented tribal primate, we are inclined to rally 'round a presumably powerful leader (even more than a flag) when under threat. This phenomenon has been described effectively in the aptly titled book *Political Animals: How Our Stone-Age Brain Gets in the Way of Smart Politics*.[89] We return to this theme shortly.

The rise of national populism isn't only a menace to liberal democracy because it embraces authoritarianism. It also threatens human rights, environmental protection and sustainability, social justice, the rule of law, and even the peaceful stability of the international world order. By providing an environment conducive to the rise of right-wing terrorist violence, it has permitted and often encouraged these emerging dangers. The next few pages focus, however, not on the threats posed *by* national populism—real though they appear to be—but on the factors responsible for its emergence and growth, nearly all of which involve an accumulation of perceived threats *to* those who, perceiving themselves threatened, have found it attractive.

The former Soviet Union was a textbook example of what might be called a "threat-based society." Very young children were exposed to the threat of shaming, ostracism, and social demotion if they deviated from officially proscribed behavior. During the Stalin terror—and to a lesser extent under Putin—adults lived under constant threat of summary judgment, imprisonment, and even execution. Their society as a whole suffered a kind of posttraumatic stress disorder as a result of its painful history of being invaded, notably by Napoleon and Hitler, but also by Poland, Lithuania, the Mongols, and others.

The dissolution of the USSR, followed by the expansion of NATO right up to Russia's borders, further exacerbated this widespread sense of being threatened, not only from within but also from the outside. By contrast with the Soviet/Russian experience of a threat-based society, up until recently the United States has seen itself as largely opportunity based, with the assumption that each generation will do better than the one that preceded it. It can be argued that the US, too, has become increasingly threat based—and thus threat obsessed—and then, because of its responses, threat endangered and diminished.

There is no single, overriding cause of national populism's surge, just as there is no single demographic or behavioral type of national populist supporter. Although much has been made of the particular attraction experienced by lower income, less-educated, blue-collar white workers, many—albeit not a majority—of Trump and Brexit supporters were middle income, and also well educated. Trump was also favored by a large proportion of quite affluent Republican voters, not all of whom are old white men or who considered themselves out of the political or ideological mainstream, whereas Brexit voters—which included a majority of working-class individuals, as expected—also comprised nearly one third of the United Kingdom's minority population. Nonetheless, there are some general patterns that can be identified, all of which speak to a block of citizens that felt (and still feel) threatened and besieged: economically, socially, demographically, even intellectually and theologically.

In *The True Believer*, an influential book published in 1951, Eric Hoffer pointed out a key (perhaps *the* key) to the success of demagogues: the existence of a disaffected populace, whose alienation from their government derives from the corrosive feeling that they are worse off than they had been—specifically, that they have lost power that they previously enjoyed. The successful demagogue condenses this dissatisfaction by messaging that he will make their country great again—which translates to "make their own situation what it had been" (or what they fantasize that it was). Demagogues seem to know intuitively that successful demagoguery is greatly facilitated if it channels resentment against convenient scapegoats, as long as these are a relatively powerless minority.

Politics, sometimes defined as the art of the possible, is often the art of manipulating threats as well, something especially applicable when it comes to national populism. Donald Trump, for all his intellectual limitations, was adroit at playing on—and in the process, enhancing—anxieties and angers that had already existed among his "base," albeit often below the surface. For the most part, prior to his candidacy, these sentiments were widely considered too embarrassing, bigoted, unjustifiable, or in other ways socially inappropriate to express publicly.

Among those threats felt at a gut level, especially by many in the white working-class electorate, racial fears are prominent—views that had not been articulated for many decades by any mainstream politician. Research has uncovered, for example, that in the 2016 presidential election, racial resentment and the threat of having lost status to a perceived inferior group activated economic and social anger and anxiety, rather than the other way around.[90] It is hard enough—but not *that* hard—to complain about one's economic situation, more difficult yet in polite company to give one's devil a microphone and heap racist animosity on others, blaming them either directly or by implication for one's circumstance.

By publicly connecting bigotry to the imprimatur of the presidency, Trump gave legitimacy to these feelings, while emboldening the expression of their most vicious, thuggish manifestations among racists, neo-Nazis, and genuine Nazis whose voices (and actions) had previously been less frequent and more veiled.[xvii]

It may seem oxymoronic, but national populism has also been globalized, benefiting not only from the Internet—with far-right websites such as 4chan and 8chan that serve as a megaphone for extremist voices—but also the public pronouncements of self-proclaimed nationalists such as Trump, whose provocative statements provide both cover and inspiration for actual violence by people who already feel marginalized and under threat. The Russian government has done its part, abetting right-wing nationalist extremism in the United States in other countries, although these interventions have consistently been denied.

The following is one example of multinational synergy. In February 18, 2017, President Trump told supporters at a rally that immigrants had caused violence in Sweden: "You look at what's happening last night in Sweden. Sweden! Who would believe this? Sweden! They took in large numbers. They're having problems like they never thought possible." But there had been no such troubles in Sweden; Trump was repeating and elaborating

---

[xvii] This book intentionally refrains from elaborating on the coronavirus pandemic, which, although an immense threat, is closer to a natural disaster than an example of personal or society-based intimidation. But when it comes to demagogic politicians fanning fear of "others," Trump's labeling COVID-19 a disease caused by the "Chinese virus," we see a textbook example of exactly that. And as this book is being written, the United States has experienced a predictable upsurge of personal vituperation directed at Chinese-Americans.

upon an inaccurate film he had seen on *Fox News* that sought to picture that country as victimized by its open-immigration policies. Then, two days later and just as Swedish commentators were deriding the US president as an uninformed laughingstock, there was in fact a violent episode in which masked youths attacked some Swedish police. It turns out that Russian government television crews had promptly arrived in the town of Rinkeby, where they paid young men to engage in a civic disturbance, apparently in an effort to validate Trump's disinformation and incitement while also stoking fearful nation-alist sentiment aimed at disrupting Swedish society,[91] expecting that many Swedes would falsely blame the staged violence on immigrants.

Here are some data for events within the United States[92] : the Southern Poverty Law Center reports a dramatic increase in the number of white nationalist groups—from 100 chapters in 2017 to 148 in 2018. The Anti-Defamation League reports a 182% increase in the distribution of white supremacist propaganda, and an increase in the number of rallies and demonstrations by white supremacy groups from 76 in 2017 to 91 in 2018. A study by the Center for Strategic and International Studies noted the number of terrorist attacks by far-right perpetrators quadrupled in the US between 2016 and 2017, and that far-right attacks in Europe rose forty-three percent during the same period.

According to the FBI, within the United States,

> there were 7,175 hate crime incidents in 2017, a 17% increase from 2016 . . . .
> The number of incidents in 2017 was also the highest yearly total since 2008.
> About 58% of the hate crimes in 2017 were motivated by race/ethnicity/ances-
> try. Digging deeper into the numbers, anti-black or African American hate
> crime rose 16% to 2,013 incidents in 2017; anti-Hispanic incidents rose 24%,
> with 427 incidents; anti-Arab crimes doubled to 102 incidents. Anti-Jewish
> hate crime incidents also rose 37% to 938 in 2017 . . . .[93]

It has become a commonplace observation that whites in the United States—and increasingly in Europe as well—feel threatened by the prospect of soon becoming a minority in "their own" countries (as though such countries are owned by their pink-skinned inhabitants), a fear exacerbated by a recent influx of refugees: mostly Islamic in Europe and mostly Latino in the US. This sense of being besieged has been enhanced by the nature of these would-be immigrants, not only their skin color, but also by their language, cultural traditions, and, in the case of Muslims, their religion—with threatening aspects of the latter magnified yet more by the existence of violent Islamic fundamentalism elsewhere in the world. Post-9/11, Americans estimated that the proportion of Muslims living in the United States was seventeen percent; the real proportion remains at barely one percent.[94]

Even among Americans not unduly anxious about diversity in their midst, many are deeply suspicious whether people of different ethnicities will satisfactorily adjust to their new home's preexisting social and cultural patterns. For them, making America "great again" means returning it to a kind of prelapsarian ideal, when things were imagined to be, if not Edenic, at least more stable and more consistent with their sense of Christian values, as well as male dominated and socioeconomically stratified in reassuring ways.

Worry about cultural estrangement, exemplified by sexual liberality, the advent of gay marriage, falling birth rates among its "native" white population, and a decline in church attendance has not been limited to the United States and Europe. Russia's Vladimir Putin got into the act, stirring up national populism within his own country, and not just others. In 2013 he warned that the "Euro-Atlantic countries" were "denying moral principles and all traditional identities: national, cultural, religious, and even sexual."[95] Modernity has been threatening in many ways to many people, not merely— as some in the "Euro-Atlantic countries" imagine—among Islamic fundamentalists yearning for the good old days of an eleventh-century caliphate.

In addition, the Internet has brought about a pace of information flow (much of it unvalidated, if not outright incorrect) that adds to a dizzying perception that, as Hamlet put it, "the world is out of joint." Collective Judeo-Christian religious faith has been threatened, along with traditional family structures and social values such that the old verities no longer apply, replaced by dizzying disorientation. All is no longer right with the world, which is to say that things are changing, leaving large numbers confused, angry, and feeling left out—which, increasingly, they are.

Here is the opening stanza of Yeats's "The Second Coming," which he began to write just after the First World War, the Russian Revolution, and ongoing violent turmoil in what was then English-occupied Ireland. The poem has experienced its own second coming in this time of intense national populism, when previously reliable traditions have been under siege:

> Turning and turning in the widening gyre
> The falcon cannot hear the falconer;
> Things fall apart; the centre cannot hold;
> Mere anarchy is loosed upon the world,
> The blood-dimmed tide is loosed, and everywhere
> The ceremony of innocence is drowned;
> The best lack all conviction, while the worst
> Are full of passionate intensity.

Those who are attracted to right-wing, racially tinged national populism converge, interestingly, with those (such as this author) who are appalled by it, both agreeing that things have been falling apart and that the center has not been holding. For those of us "vexed to nightmare" by national populism, who worry that this particular rough beast, "its hour come 'round at last," has rushed to Washington, DC, to be born, "Trumpism" and all that it represents is a serious threat indeed.

At the same time, it warrants repeating that national populists see their movement as a *response* to a range of threats: increasingly outnumbered by growing numbers of racial and ethnic minorities plus immigrants, who, taking their jobs, seducing their children from the path of religious and ideological righteousness, comprise a "blood-dimmed tide" of strangers, weirdos, and cultural anarchists abetted by an out-of-touch intellectual and political elite who, when not looking down on them (e.g., Hillary Clinton's "basket of deplorables") have abandoned and ignored them, threatening not only their self-esteem, but also all they hold dear. Small wonder that both sides are full of passionate intensity.

It is not only ideology that is unequal on this particular playing field. Income and wealth inequality have been rising in the United States, which has the highest Gini index[xviii] (hovering around 45) in the Western world whereas Scandinavian countries vary from 29 (Denmark) to 24.9 (Sweden).[96] Not surprisingly, these latter social democracies have experienced substantially less national populist backlash. Story has it that a French politician, Alexandre Ledru-Roilin, was in Paris during the tumultuous events of 1848. When a mob poured through the streets, he cried out, "There go the people! I must follow them, for I am their leader!"

Threats to liberal democracy derive not only from fear-mongering demagogic leaders, but also from growing inequality (traditionally a concern of the political left rather than the right) and the demands of disaffected people, including a backlash from the Civil Rights movement, minority-focused identity politics, and heightened "political correctness" in the United States.

A similar sense of being under cultural and demographic attack via the European Union's open borders and consequent increased migration into the United Kingdom strongly influenced the Brexit vote of 2016. Preceding and almost certainly influencing this vote was a sustained fear-mongering campaign ("I must follow them, for I am their leader!") that exaggerated the numbers of immigrants arriving through the European Union, along with the social welfare costs they allegedly impose—in summary, the threat they pose to the "British way of life." Worldwide, national populism has become indistinguishable from nativist populism.

[xviii] This is a statistical measure of inequality across a population, with the highest (one hundred) obtained when inequality is total—i.e., one person has everything—and the lowest (zero) when everyone is equal.

Add to this the perception of patronizing superiority by urbane, cosmopolitan elites (aka limousine liberals) who have, to some extent, ignored national identity as well as the plight of "normal people" close to home in favor of concern for others far away. This fear of and anger toward global elites and cosmopolitans is not altogether new, shown by a satirized character in Charles Dickens's *Bleak House*. Mrs. Jellyby is a philanthropist obsessed with bettering the lot of children thousands of miles away while neglecting the needy in her own neighborhood and even her own offspring. As described in that mid nineteenth-century novel, "her eyes had a curious habit of seeming to look a long way off, as though they could see nothing nearer than Africa."[xix]

For many years, such indifference on the part of the intelligentsia to needs "at home" was not exceptionally painful, insofar as working-class whites in the United States and elsewhere experienced reliably improving economic conditions, leading to the realistic expectation that their children would do better than themselves, while at the same time enjoying the satisfaction of knowing that their social situation was consistently better than that of ethnic and racial minorities, and that their country was largely defined—at least in their own mind—as an extension of themselves. But as this privileged status began to wane, so did their confidence and satisfaction.

What sociologists call "achieved identity" became less gratifying as it became less available, replaced more and more by "ascribed identity," such as ethnicity, skin color, and religious affiliation. And then, to make matters worse, ascribed identity itself became increasingly threatened as a result of demographic trends, including immigration along with the election of a black president who tried, especially during the first half of his administration, to minimize being seen as a threatening angry black man by persistently seeking compromise with Republicans. (It didn't work.)

Resistance to loss of ascribed identity doesn't even require that the interlopers be of a different religion, ethnic group, or, indeed, very different at all. Germans living in what had been West Germany were in no significant way distinguishable from those in East Germany, and yet it was reported that when the Berlin Wall fell and West and East were reunited into one Germany, those from the East began chanting, "*Wir sind ein Volk!*" ["We are one people!"] to which the West Germans responded, "*Wir auch!*" ["So are we!"]. Just as under communism some people were "more equal" than others, some people—perhaps most—see their immediate tribe as more equal, or at least more familiar and therefore more acceptable than others. Even when the tribal differences are so trivial as to be essentially nonexistent.

Francis Fukuyama has argued that the plaint of "I want my country back" is less one of aggrieved nationalism per se than of threatened loss of identity. His book *Identity* is summarized in its subtitle: *The Demand for Dignity and the Politics of Resentment*.[97]

[xix] I thank Professor Kwame Anthony Appiah, who brought Mrs. Jellyby to my attention.

White identity crisis in the United States resonated with President Trump's call to build a wall along the border with Mexico, thereby supposedly protecting the homeland from Latin American migrants. And in the United Kingdom, a vote—just a few months before Trump was elected—to leave the European Union, thereby supposedly protected the British homeland from Eastern European and Islamic immigrants.

In each case (mirrored in various locally distinct ways in other countries where right-wing national populism has advanced), "America first," "Britain for the British," or the equivalent has been the rallying cry, whether overt or covert, manifesting a component of what Robin DiAngelo has called "white fragility,"[98] especially characterized by guilt, anger, and fear.[xx] And this, in turn, has led to an embrace—in Europe no less than in the United States—of leaders who promise to control the flood of immigrants.[99] It can be seen, for example, in votes for the Vox political party in Spain, Alternative for Germany in (unsurprisingly) Germany, the Five Star Movement in Italy, and the UK Independence Party in the United Kingdom.

The Nation columnist Kai Wright writes:

> Hell hath no fury like a white man scorned. Donald Trump Jr. says he's worried for his multimillionaire sons, that it's a scary time to be a (white) man. . . . [A]n array of Republican senators repeated this odd concern. At first, I marveled at how little it takes to make a powerful white man feel like he's in danger. But then I realized: They're correct—we absolutely are a threat to them.
>
> They've looked around and rightly noticed how many of us do not draw our power from proximity to them. In the Obama era, they watched the Dreamers discard the white man's idea of citizenship and demand a fundamentally new conversation about immigrant rights. They watched black people build a movement on an irrefutable statement of self-worth, one that requires no white person's approval to be true and potent. And now they are watching as millions of women refuse to carry the shame of their male predators. So no wonder these white men thrash and howl with defensive rage. Good. Let them be afraid. Because it's true: We are coming for them, and for their power, too.[100]

No wonder many blue-collar whites—with their newfound sense of fragility, increasingly alienated by the Civil Rights movement, as well as left out of identity politics—have felt deeply threatened, despite the fact that they long benefited, knowingly or not, from a system of advantage. As Wright points out, some of that threat is genuine—and overdue. (Those claiming that class trumps race in this regard might want to compare

---

[xx] Her argument is actually more subtle than this, notably including the defensiveness with which many whites respond when confronted with evidence of their own implicit racial bias.

the obstacles faced by poor whites in the United States with those confronted by poor blacks.) Writing in *The New Yorker* shortly after Donald Trump's election in 2016, Toni Morrison—as usual—put her finger on it, succinctly and perceptively: "So scary are the consequences of a collapse of white privilege that many Americans have flocked to a political platform that supports and translates violence against the defenseless as strength. These people are not so much angry as terrified . . . ."[101]

Not all Trump voters were terrified or irredeemably bigoted; a significant number had been Obama voters who switched to Trump but then back in the 2018 election. Research has shown that the feeling of being under siege on the part of many whites, to some extent independent of socioeconomic circumstance, is enhanced in proportion as social, economic, and demographic changes have been rapid—a sense of threat that has been especially pronounced when it comes to attitudes toward immigrants. The problem of threatened white identity has not only involved feeling stuck while others—especially people of color—seem to be moving forward, but also it is a reaction to other people moving "in."[102]

One of many studies of white working-class attitudes carried out in the aftermath of the 2016 presidential election reported the following observations by a resident of the small town of Dundalk, Maryland:

> I do not deny that there is very real, longstanding racism. [But] if you want to know why communities like Dundalk [Maryland] voted for Trump, it's not really bigotry in itself; it's fear, it's the sense of alienation, it's the sense of desperation, it's the sense of a lack of answers. . . .
>
> If you feel like you've got a place in the society around you and your own situation is not tottering on the brink, you're secure enough to open the door to other people, literally and figuratively. On the other hand, if you're fearful, desperate, alienated, you start looking for ways to be suspicious of other people.[103]

Based on interviews such as this, Professor Andrew Cherlin of Johns Hopkins University went on to conclude that Donald Trump's masterstroke had been

> to recognize the desperation of the white working class over the deteriorating industrial economy and to encourage their tendency to racialize that desperation. Neither economics nor identity politics can be said to be the more important factor. Perhaps one without the other—economics in a setting where no one racialized it, or racial prejudice at a time of economic prosperity—would not have brought about the same result. Together, they were tinder for the bonfire that resulted. And Trump was the match.

It seems paradoxical for the historical beneficiaries of the US economy to feel themselves marginalized, but as manufacturing jobs have increasingly moved overseas, to be replaced by an epidemic of opioid addiction and deaths, along with a general waning of privileged status, such threats multiplied. Financial inequality skyrocketed as the share of US national income obtained by the top one percent, calculated as their percentage of national income, more than doubled from 1980 to 2015. During those same years, the share obtained by the lowest fifty percent has declined steadily and painfully.[104]

Research by Roberto Foa and Yascha Mounk has emphasized the threat that inequality poses for democracy itself: the greater the inequality, the lower is citizen confidence in basic democratic institutions.[105] Threats of losing status relative to fellow society members, combined with loss of measurable wealth and growing fear about the future—including racial and demographic anxieties—can thus contribute to the undermining of democracy itself, which in turn generates a turn toward authoritarianism, and the appeal of posturing "strong-man" leaders.

An article appearing in *The New York Times* on November 4, 2018, shortly before the US mid-term elections, was titled "At Trump's Rallies, Women See a Hero Protecting Their Way of Life." Here is a sample:

Standing in an airplane hangar in the mid-autumn chill awaiting the arrival of President Trump, Joan Philpott said she was angry and scared. Only Mr. Trump, she said, can solve the problems she worries most about. "He wants to protect this country, and he wants to keep it safe, and he wants to keep it free of invaders and the caravan and everything else that's going on," said Ms. Philpott, 69, a retired respiratory therapist. Ms. Philpott was one of thousands of women who braved a drizzle for hours to have the chance to cheer Mr. Trump at a rally here on Thursday. While political strategists and public opinion experts agree that Mr. Trump's greatest electoral weakness is among female voters, here in Columbia and places like it, the president enjoys a hero-like status among women who say he is fighting to preserve a way of life threatened by an increasingly liberal Democratic Party.

'He understands why we're angry," Ms. Philpott said, "and he wants to fix it." As Republican candidates battle to keep their congressional majorities in the midterm elections on Tuesday, Mr. Trump is crisscrossing the country to deliver a closing argument meant to acknowledge—and in many cases stoke—women's anxieties. At rally after rally, he has said that women "want security," warning of encroaching immigrants, rising crime and a looming economic downturn if Democrats gain power.

The underlying psychology at work is likely very old—part of our primate evolutionary history. Evolutionary psychologist Hector Garcia has pioneered the assessment that monotheism itself is a derivative of *Homo sapiens*' deep-seated tendency to defer to an alpha male, thereby ensconcing God as the dominant leader, endowed with such predictable qualities as large size, great power, intolerance of competitors, sexual jealousy, and the supposed ability to provide benefits to His faithful followers.[106]

Garcia revisited this territory in his book *Sex, Power and Partisanship*, showing how the trope of Democrats as the "mommy party" and Republicans as the "daddy party" helps make sense of our current partisan divide, while illuminating how our evolved psychology plays into the impact of threats upon political orientation. In times of stress and threat, those especially affected find themselves attracted to political leaders and platforms that promise authoritarian certainty, including a vigorous, uninhibited, and even potentially violent response to those individuals and circumstances regarded as threatening.

Psychologist Michelle Gelfand has developed a fruitful approach to understanding cross-cultural differences by distinguishing between "tight" and "loose" societies.[107] The former—such as Germany, Japan, and Singapore—are characterized by clearly articulated and rigorously enforced rules for and expectations of acceptable behavior, whereas the latter—such as Australia, Brazil, and the Netherlands—grant more behavioral leeway. Such designations can vary with the behaviors in question (Italy, for example, is generally loose, but not when it comes to food), but the pattern appears to be meaningful, and as Gelfand points out, threats tend to make societies tighter. This correlates with the appeal of leaders who promise a kind of reassuring tightness to those who feel threatened.

Human nature and its enabling hardware changes at a biological snail's pace whereas cultural circumstances gallop ahead, largely untethered from our slow-moving evolution.[108] As a result, we are stuck with a kind of zombie psychology that staggers along, frequently out of touch with current needs and situations. There was a time in our Pleistocene past when alpha males, despite their fierce despotism (and in some cases because of it), were an asset, especially when dealing with outside enemies, and maybe to quell personal disputes within the group. But that was a long time ago. Current presidents don't stand in the fighting vanguard, huffing and puffing and blowing the other guys down. Our species' susceptibility to leaders' theatrics nonetheless remains deeply entrenched, a long-ago asset now turned liability that is all the more evident when politicians effectively stoke our fears, exaggerating the seeming threats, all the while thumping their chests and proclaiming themselves the best, the most, and maybe even the only genuine alpha creatures this side of God.

Whether they are is debatable. Undeniable, however, is that, when they gain power, such leaders often use threats to underpin their newfound authority. For one, they are frequently inclined to manipulate their followers by exaggerating or, if necessary, creating an external danger, thereby hoping to benefit by the ensuing "rally 'round the flag" effect.[109] In Shakespeare's play *Henry IV, Part 2*, the dying king advises his son to "busy giddy minds with foreign quarrels," thereby consolidating support at home.

As the American Civil War loomed, Secretary of State William Seward advised President Abraham Lincoln to consider declaring war against France or Spain as a way to unite the country. In 1983, almost immediately after two truck bombs destroyed a marine barracks in Beirut, killing 241 US service personnel and also constituting a huge political embarrassment, President Ronald Reagan announced a military operation against the tiny Caribbean country of Grenada, claiming that it had become a threat to American medical students on that island. Such tactics—concocting a war to distract attention from other problems—were satirized in the motion picture *Wag the Dog*.

Regardless of whether they obtained their position because their citizens perceived themselves under threat and flocked to them in search of a protecting alpha individual, rulers have also historically used brutal tactics against the population they seek to dominate, secondarily to eliminate their immediate opponents, but—at a deeper level—expecting that by threatening to do the same thing to other would-be enemies, they will inhibit further resistance. Sometimes it has worked, but typically not for long. And more often, the threat of visiting more pain and suffering on a subjected population has eventually backfired. Edward Gibbon wrote that the Roman Emperor Maximin's "wanton and ill-timed cruelty, instead of striking terror, inspired hatred" of the sort that has often led to greater opposition to such rulers, who, seeking to reduce the threat to their regime, have ended up increasing it. Seneca—who served as political adviser to Roman leaders, including an ultimately fatal association with Nero—warned in his treatise *Of Clemency* (intended as advice to Nero), "A cruel king increases the number of his enemies by destroying them; because the parents and children of those who are put to death, and their relatives and friends, step into the place of each victim."

Henry VIII ordered the execution of numerous heretics during his rule. According to philosopher David Hume in his *History of England* (written with Tobias Smollett), "those severe executions, which in another disposition of men's minds would have sufficed to suppress [their heresy] now served only to diffuse it the more among the people, and to inspire them with horror against the unrelenting persecutors."

The potency of such counterreactions has long been known and sometimes used to advantage by those being persecuted, even to the point of encouraging a heavy and threatening hand so as to generate resistance. Anti-Nazi partisans in occupied France murdered German officers and officials, not so much *despite* knowing that the immediate result would be murderous reprisals against the French citizenry, but rather betting that, in the longer term, such brutality would generate more anti-German sentiment.

This so-called "tyranny effect"—the boomerang consequence of attempting to rule by violent threats made manifest—generally worked.

Seneca's insight was reinforced nearly two millennia later by the American Vietnam War officer and critic John Paul Vann, who was outraged—not so much morally as from his hard-headed "realist" perspective—by the US policy of indiscriminately bombing and shelling Vietnamese noncombatants, noting that instead of reducing the insurgent threat, such behavior "kills many, many more civilians than it ever does VC [Viet Cong] and as a result, makes more VC."[110]

There do not appear to be animal parallels. Dominant males in particular typically threaten would-be challenger males and strive to keep their harems in line by threatening females who stray or interact with strange males. Although such despots are eventually overthrown, nearly always to be replaced by new ones, such local revolutions seem to be inspired purely by individual males following their own competitive, Darwinian logic, which induces them to maximize their personal fitness opportunities, rather than because the violence perpetrated by harem masters leads others to revolt.

Whatever the explanatory power of a biological perspective when it comes to the appeal of national populism, it seems clear that being threatened is a key (albeit probably not the only one) to understanding that appeal. Given the harm liable to result from Brexit, along with the Donald Trump's exacerbation of the plight of those who voted for him, progressives have been especially perplexed by populism's popularity, because these people—the populace, less wealthy than their upper income fellows—are traditionally *their* base, not that of the political right.

Simultaneously, the Republican Party has been increasingly running scared about the reliability of *its* base, but only in part because of defections, as seen in the 2018 congressional US elections. Its brand of Trumpist populism threatens the Party because of its growing unpopularity among young voters and also by the specter of being outnumbered as a result of demographic changes. Under these combined threats, Republicans have turned, quite brazenly, to voter suppression of different sorts: closing polling places in minority neighborhoods, insisting on Draconian voter registration procedures, opposing the franchise for ex-convicts who have served their terms and thus ostensibly fulfilled their debt to society, and much more.

The outcome, as with so many other responses we have seen to threats—and as with so many more we shall yet encounter—has been to increase the overall threat not only to the country's democratic traditions, but also to the GOP itself, insofar as it is liable to be seen as not only an elderly, white, male, and therefore minority party, but also one that has set itself against the better angels of America's nature.

As of 2019, the domestic costs of US national populism have largely been contained short of obvious mass tragedy, although it is abundantly clear—except, ironically, to many of those most injured—that the psychology and politics of racial and ethnic resentment has actually victimized white America, just as it has long penalized people of color. National populism in general and "Trumpism" in particular appeal, counterintuitively, to those who find themselves besieged by precisely those policies that Trump and other national populists champion: a less reliable social safety net, reduced social opportunities, a more damaged natural environment, and an increased prospect of wars, both economic and military. By promising to discriminate against the discriminated against, national populism implicitly promises to benefit its base whereas, in practice, it disadvantages them even more.

For example, labor unions have historically been especially involved in overseeing worker safety. In recent decades, this role has been largely taken up by the federal government, particularly the Occupational Safety and Health Administration. But, under the Trump Administration, things changed. The following account was part of an editorial in *The New York Times* arguing that the Trump Administration, although elected in part because of enthusiasm on the part of low-wage, blue-collar workers, has in fact increased their difficult circumstances rather than improving their lot:

> On Sept. 15, a worker at a Peco Foods chicken processing plant in Pocahontas, Ark., lost his left ring finger in an industrial bagging machine. OSHA did not send an inspector to the plant after the accident. The next month, the government gave the plant permission to increase the speed at which chickens are processed. On Dec. 27, another employee lost a finger—this time, his right index finger. Once again, OSHA did not send an inspector. These decisions, and the administration's broader pattern of actions and inaction, is sending a clear message to American workers: You're on your own.[111]

Efforts to restore the expectations of white superiority, powered in large part by the mistaken perception that government programs disproportionately benefit non-whites, have particularly caused economic and social havoc within red states, contributing, for example, to reduced education budgets in Kansas, refusal to accept Medicaid expansion under the Affordable Care Act in Texas (and nearly all of the conservative "solid South"), and support for pro-gun legislation in Missouri. These populist backlashes have, in turn, reduced economic progress and raised dropout rates in Kansas, heightened opioid abuse in Texas, and increased the suicidal use of firearms—especially by whites—in Missouri. In short, right-wing threat-powered populist and racist policies have, to a large extent, backfired and most harmed those who embrace them.[112]

It has long been known that racism had a devastating effect on people of color in the United States. Now it is increasingly clear that policies and politicians embraced by resentful and marginalized whites in an effort to restore their privilege and make them "great again" have instead made them lesser: more unhappy, unhealthy, and prematurely dead.

And yet, the potential exists for resolving shared threats in nontragic, or at least nonviolent ways. The late Israeli novelist and peace activist Amos Oz made the following observation, the implications of which go beyond his long-time concern of seeking accommodation in the Middle East:

> The Israeli-Palestinian conflict is a clash of right and right. Tragedies are resolved in one of two ways: The Shakespearean way or the Anton Chekhov way. In a tragedy by Shakespeare, the stage at the end is littered with dead bodies. In a tragedy by Chekhov, everyone is unhappy, bitter, disillusioned and melancholy, but they are alive. My colleagues in the peace movement and I are working for a Chekhovian, not a Shakespearean conclusion.[113]

NOTES

1. Grogan-Kaylor, A., Ma, J., and Graham-Bermann, S.A. 2018. The case against physical punishment. *Current Opinion in Psychology, 19*, 22–27.
2. Saville, G. 2009. *The Complete Works of George Saville, First Marquess of Halifax*. Ithaca, N.Y.: Cornell University Press.
3. https://news.gallup.com/poll/1606/death-penalty.aspx.
4. https://news.gallup.com/poll/1690/religion.aspx.
5. Wrangham, R. 2019. *The Goodness Paradox*. New York, N.Y.: Pantheon.
6. Kristof, N. 2019. When we kill. *The New York Times*. https://nyti.ms/2KpJWHJ.
7. Espy, M.W. and Smykla, J.O. 2004. *Executions in the United States, 1608–2002: The ESPY File*. Ann Arbor, Mich.: Inter-university Consortium for Political and Social Research.
8. https://deathpenaltyinfo.org/race-and-death-penalty.
9. https://deathpenaltyinfo.org/race-and-death-penalty.
10. Edens, J.F., Davis, K.M., Fernandez Smith, K., and Guy, L.S. 2013. No sympathy for the devil: attributing psychopathic traits to capital murderers also predicts support for executing them. *Personality Disorders: Theory, Research, and Treatment, 4*(2), 175–181.
11. https://deathpenaltyinfo.org/costs-death-penalty
12. https://theconversation.com/theres-no-evidence-that-death-penalty-is-a-deterrent-against-crime-43227.
13. Zimring, F.E. and Hawkins, G.J. 1973. *Deterrence: The Legal Threat in Crime Control*. Chicago, Ill.: University of Chicago Press.
14. Zimring, F.E. 2003. *The Contradictions of American Capital Punishment*. New York, N.Y.: Oxford University Press.
15. https://deathpenaltyinfo.org/law-enforcement-views-deterrence.
16. https://deathpenaltycurriculum.org/student/c/about/arguments/argument1b.htm.
17. Barash, D.P. and Lipton, J.E. 2010. *Payback: Why We Retaliate, Redirect Aggression and Take Revenge*. New York, N.Y.: Oxford University Press.
18. Radelet, M.L. and Akers, R.L. 1996. Deterrence and the death penalty: the views of the experts. *Journal of Criminal Law & Criminology, 87*, 1–16.

19. Godfrey, M.J. and Schiraldi, V. 1995. *How Have Homicide Rates Been Affected by California's Death Penalty?* San Francisco, Calif.: Center on Juvenile and Criminal Justice.

20. Donohue, J.J. 2009. Estimating the impact of the death penalty on murder. UC Berkeley: Berkeley Program in Law and Economics. https://escholarship.org/uc/item/1gk0r77m.

21. Bowers, W.J. and Pierce, G.L. 1980. Deterrence or brutalization: what is the effect of executions? *Crime & Delinquency, 26*(4), 453–484.

22. Ehrlich, I. 1975. The deterrent effect of capital punishment: a question of life and death. *American Economic Review, 65* (3), 397–417.

23. Donohue, J.J., III and Wolfers, J. 2009. Estimating the impact of the death penalty on murder. *American Law and Economics Review, 11*(2), 249–309.

24. National Research Council. 2012. *Deterrence and the Death Penalty.* Washington, D.C.: National Academies Press.

25. Donohue, J.J. and Wolfers, J. 2006. Uses and abuses of empirical evidence in the death penalty debate. *Stanford Law Review 58,* 791–846.

26. Bailey W. and Peterson, R. 1994. Murder, capital punishment, and deterrence: a review of the evidence and an examination of police killings. *Journal of Social Issues 53,* 50–71.

27. Federal Bureau of Investigation. 1998. *Law Enforcement Officers Killed and Assaulted, 1998.* Washington, D.C.: Uniform Crime Reports.

28. Donohue, J.J. 2015. There's no evidence that death penalty is a deterrent against crime. theconversation.com.

29. Zimring, F.E., Fagan, J., and Johnson, D.T. 2009. Executions, deterrence and homicide: a tale of two cities. Columbia Public Law School research paper no. 09-206, CELS 2009 4th Annual Conference on Empirical Legal Studies Paper. https://ssrn.com/abstract=1436993 or http://dx.doi.org/10.2139/ssrn.1436993.

30. https://deathpenaltyinfo.org/executions-year.

31. Donohue, n 20.

32. Note no. 33/2019, sent from Dato Erywann Pehin Yusof, Minister of Foreign Affairs, Brunei Darussalam to the UN High Commissioner for Human Rights, April 8, 2019.

33. Blum, D. 2010. *The Poisoner's Handbook: Murder and the Birth of Forensic Medicine in Jazz Age New York.* New York, N.Y.: Penguin

34. Inglehart, R. 2018. *Cultural Evolution.* New York, N.Y.: Cambridge University Press.

35. Chen, Y. and Koenig, H. 2006. Traumatic stress and religion: is there a relationship? A review of empirical findings. *Journal of Religion and Health, 45*(3), 371–381.

36. Shaw, A. Joseph, S., and Linley, P.A. 2005. Religion, spirituality, and posttraumatic growth: a systematic review. *Mental Health, Religion & Culture, 8*(1), 1–11.

37. Ellison, C.G. 1991. Religious involvement and subjective well-being. *Journal of Health and Social Behavior, 32*(1): 80–99.

38. Henrich, J., Bauer, M., Cassar, A., Chytilová, J., & Purzycki, B.G. 2019. War increases religiosity. *Nature Human Behaviour, 3,* 129–135.

39. Cesur, R., Freidman, T., & Sabia, J.J. 2018. *Death, Trauma and God: The Effect of Military Deployments on Religiosity.* Cambridge, Mass.: National Bureau of Economic Research.

40. Astore, W. 2010. "In place of mental health care, are some troops being evangelized?" *HuffingtonPost.*http://www.huffingrangetonpost.com/william-astore/in-place-of-mental-health_b_677602.html.

41. Dreher, R. 2003. Ministers of war. *National Review.* http://www.nationalreview.com/article/210615/ministers-war-rod-dreher.

42. Jacoby, S. 2016. *Strange Gods: A Secular History of Conversions.* New York, N.Y.: Vintage.

43. Zapata, O. 2018. Turning to God in tough times? Human versus material losses from climate disasters in Canada. *Economics of Disasters and Climate Change, 3,* 259–281.

44. Miller, M. 2018. *Circe.* New York, N.Y.: Little, Brown.

45. Falwell: blame abortionists, feminists and gays. https://www.theguardian.com/world/2001/sep/19/september11.usa9.

46. Baker, J. 2008. Who believes in religious evil? An investigation of sociological patterns of belief in Satan, hell, and demons. *Review of Religious Research, 50*(2): 206–220.

47. http://www.thearda.com/Archive/Files/Codebooks/FTHMAT11_CB.asp#V142.

48. https://www.pewforum.org/2018/08/29/the-religious-typology/#group-profiles-the-highly-religious.

49. Greeley, A.M. 1989. *Religious Change in America.* Cambridge, Mass.: Harvard University Press.

50. Shariff, A.F. and Rhemtulla, M. 2012. Divergent effects of beliefs in heaven and hell on national crime rates. *PLoS One, 7,* e39048.

51. Barro, R. and McCleary, R. 2003. Religion and economic growth across countries. *American Sociological Review, 68,* 760–781.

52. Shariff, A.F. and Aknin, L.B. 2014. The emotional toll of hell: cross-national and experimental evidence for the negative well-being effects of hell beliefs. *PLoS One, 9*(1), e85251.

53. Greenblatt, S. 2018. Damn it all. *The New York Review of Books.* December 20.

54. Ibid.

55. Nesse, R. 2019. *Good Reasons for Bad Feelings: Insights from the Frontier of Evolutionary Psychiatry.* New York, N.Y.: Dutton.

56. Laderman, G. 2003. *Rest in Peace: A Cultural History of Death and the Funeral Home in Twentieth-Century America.* New York, N.Y.: Oxford University Press.

57. Greenberg, J., Solomon, S., and Pyszczynski, T. 2016. *The Worm at the Core: On the Role of in Life.* New York, N.Y.: Penguin.

58. Burke, B.L., Martens, A., and Faucher, E.H. 2010. Two decades of terror management theory: a meta-analysis of mortality salience research. *Personality and Social Psychology Review, 14*(2), 155–195.

59. Kristof, N. 2019. *The New York Times,* March 20, https://nyti.ms/2Cx9UTO.

60. https://www.pewresearch.org/fact-tank/2017/06/22/key-takeaways-on-americans-views-of-guns-and-gun-ownership/.

61. https://www.huffpost.com/entry/crime-rates-donald-trump_n_57a8aa11e4b056bad2164226.

62. https://www.nybooks.com/articles/2018/04/05/guns-bang-for-the-buck/.

63. http://fortune.com/2018/02/20/australia-gun-control-success/.

64. 2012. http://aler.oxfordjournals.org.

65. https://www.ncbi.nlm.nih.gov/pmc/articles/PMC4504296/.

66. https://www.nytimes.com/2015/12/22/health/in-missouri-fewer-gun-restrictions-and-more-gun-killings.html.

67. Hemenway, D. 2017. *Private Guns, Public Health.* Ann Arbor, Mich.: University of Michigan Press.

68. https://www.pewresearch.org/fact-tank/2019/10/16/share-of-americans-who-favor-stricter-gun-laws-has-increased-since-2017/.

69. https://vizhub.healthdata.org/gbd-compare/#.

70. Lankford, A. 2016. Public mass shooters and firearms: a cross-national study of 171 countries. *Violence and Victims, 31*(2), 187–199.

71. Zimring, F.E. and Hawkins, G. 1999. *Crime Is Not the Problem: Lethal Violence in America.* New York, N.Y.: Oxford University Press.

72. Peeples, L. 2019. What the data say about police shootings. *Nature, 573,* 24–26.

73. Miller, M., Hemenway, D., and Azrael, D. 2007. State-level homicide victimization rates in the US in relation to survey measures of household firearm ownership, 2001–2003. *Social Science & Medicine, 64*(3), 656–664.

74. https://www.tsa.gov/blog/2019/02/07/tsa-year-review-record-setting-2018.

75. Cunningham, R.M., Walton, M.A., and Carter, P.M. 2018. The major causes of death in children and adolescents in the United States. *New England Journal of Medicine, 379*(25), 2468–2475.

76. Lubin, G., Werbeloff, N., Halperin, D., Shmushkevitch, M., Weiser, M., and Knobler, H.Y. 2010. Decrease in suicide rates after a change of policy reducing access to firearms in adolescents: a naturalistic epidemiological study. *Suicide and Life-threatening Behavior, 40*(5), 421–424.

77. Hemenway, D., Shinoda-Tagawa, T., and Miller, M. 2002. Firearm availability and female homicide victimization rates among 25 populous high-income countries. *Journal of the American Medical Women's Association, 57*(2), 100–104.

78. Sorenson, S.B. and Wiebe, D.J. 2004. Weapons in the lives of battered women. *American Journal of Public Health, 94*(8), 1412–1417.

79. Hemenway, D. 2017. *Private Guns, Public Health.* Ann Arbor, Mich.: University of Michigan Press.

80. https://www.cdc.gov/nchs/data/nvsr/nvsr63/nvsr63_09.pdf.

81. Metzl, J.M. and MacLeish, K.T. 2015. Mental illness, mass shootings, and the politics of American firearms. *American Journal of Public Health, 105*(2), 240–249.

82. https://www.nytimes.com/2017/11/07/world/americas/mass-shootings-us-international.html.

83. Stone, M.H. 2015. Mass murder, mental illness, and men. *Violence and Gender, 2*(1), 51–86.

84. Swanson, J.W., Holzer, C.E., III, Ganju, V.K., and Jono, R.T. 1990. Violence and psychiatric disorder in the community: evidence from the Epidemiologic Catchment Area surveys. *Psychiatric Services, 41*(7), 761–770.

85. Kalesan, B., Vasan, S., Mobily, M.E., et al. 2014. State-specific, racial and ethnic heterogeneity in trends of firearm-related fatality rates in the USA from 2000 to 2010. *BMJ Open, 4,* doi: 10.1136/bmjopen-2014-005628.

86. Bazelon, E. 2019. *Charged: The New Movement to Transform American Prosecution and End Mass Incarceration.* New York, N.Y.: Random House.

87. Klarevas, L., 2016. *Rampage Nation: Securing America from Mass Shootings.* Amherst, N.Y.: Prometheus Books.

88. Donohue, J. and Boulouta, T. 2019. The assault weapons ban worked. *The New York Times.* September 5 https://www.nytimes.com/2019/09/04/opinion/assault-weapon-ban.html.

89. Shenkman, R. 2016. *Political Animals: How Our Stone-Age Brain Gets in the Way of Smart Politics.* New York, N.Y.: Basic Books.

90. Sides, J., Tesler, M., and Vavreck, L. 2017. The 2016 U.S. election: how Trump lost and won. *Journal of Democracy, 28,* 24–34.

91. https://www.nytimes.com/2019/08/10/world/europe/sweden-immigration-nationalism.html.

92. https://www.factcheck.org/2019/03/the-facts-on-white-nationalism/.

93. https://ucr.fbi.gov/hate-crime/2017/topic-pages/tables/table-1.xls.

94. Duffy, B. 2019. *Why We're Wrong About Nearly Everything.* New York, N.Y.: Basic Books.

95. Quoted in https://www.theatlantic.com/magazine/archive/2017/03/its-putins-world/513848/.

96. https://www.cia.gov/library/publications/the-world-factbook/rankorder/2172rank. html.

97. Fukuyama, F. 2018. *Identify: The Demand for Dignity and the Politics of Resentment.* New York, N.Y.: Farrar, Straus and Giroux.

98. DiAngelo, R. 2018. *White Fragility.* New York, N.Y.: Beacon Press.

99. Kaufmann, E. 2019. *Whiteshift: Populism, Immigration, and the Future of White Majorities.* New York, N.Y.: Henry Abrams.

100. Wright, K. 2018. White men have good reason to be scared. *The Nation,* November 5. https://www.thenation.com/article/archive/white-men-have-good-reason-to-be-scared/.

101. https://www.newyorker.com/magazine/2016/11/21/making-america-white-again.

102. Major, B., Blodorn, A., and Blascovich, G. 2018. The threat of increasing diversity: why many white Americans support Trump in the 2016 presidential election. *Group Processes & Intergroup Relations, 21*(6), 931–940.

103. Cherlin, A.J. 2019. *In the Shadow of Sparrows Point: Racialized Labor in the White and Black Working Classes.*New York, N.Y.: Russell Sage Foundation.

104. https://wir2018.wid.world/.

105. Foa, R. and Mounk, Y. 2017. The end of the consolidation paradigm. *Journal of Democracy Web Exchange, 28,* 1–26.

106. Garcia, H. 2015. *Alpha God.* Amherst, N.Y.: Prometheus Books..

107. Gelfand, M. 2018. *Rule Makers, Rule Breakers: How Tight and Loose Cultures Wire Our World.* New York, N.Y.: Scribners.

108. Barash, D.P. 1986. *The Hare and the Tortoise: Culture, Biology and Human Nature.* New York, N.Y.: Viking.

109. Barash, D.P. 1994. *Beloved Enemies: Exploring Our Need for Opponents.* Buffalo, N.Y.: Prometheus.

110. Elster, J. 2015. *Explaining Social Behavior.* New York, N.Y.: Cambridge University Press.

111. 2019. Trump's War on Worker Rights. *The New York Times,* June 4. https://www.nytimes. com/2019/06/03/opinion/trump-worker-safety-osha.html.

112. Metzl, J.M. 2019. *Dying of Whiteness: How the Politics of Racial Resentment Is Killing America's Heartland.* New York, N.Y.: Basic Books.

113. https://www.nytimes.com/2013/01/29/opinion/global/roger-cohen-sitting-down-with-amos-oz.html.

Section 3

International Affairs

THREATS AND RESPONSES to them are clearly baked into much of the natural world and are also widely represented when it comes to personal interactions as well as the edgier structures of human society. Hence, it isn't surprising that they also loom large when it comes to international affairs. We would all like to think that when countries interact, they seek and often find mutually beneficial, win–win outcomes. Sometimes this is so, but when relationships are fraught—notably, when it comes to national security—threats are prominent, paramount, and often counterproductive as well.

CONVENTIONAL DETERRENCE

"If you want peace," goes the saying, attributed to the Roman general Vegetius, and endorsed by many in the centuries to follow, "prepare for war." The assumption is that by preparing for war, you are threatening potential opponents with a bad outcome if they attack you. Sometimes it has worked.

The Great Wall of China, parts of which were begun as long ago as the seventh century BCE and mostly built during the Ming Dynasty (1368–1644) is one of the world's great construction projects, consisting of nearly four thousand miles of actual wall, running primarily east to west, and built in parts out of stone, but also in different locales, of wood and tamped earth, plus trenches and various natural barriers. Outfitted with watchtowers, signal towers, and garrison posts, it has been claimed (incorrectly) to be visible from outer space.

The Great Wall was intended to defend the Chinese heartland against nomadic tribes to the north, including the Mongols. Although it was ultimately overcome by the Manchu people, its long-lasting success stands as a monument to what is sometimes called "deterrence by denial"—the threat that an attacker will be denied the prospect of success. The Maginot Line, constructed by France along its border with Germany prior to the Second World War, was also designed to deter by denial, but the German armies simply bypassed it and went through Belgium.

By contrast, the Roman Empire enjoyed a degree of peace (the *Pax Romana*) by relying on what has been called "deterrence by punishment"—more accurately, the *threat* of punishment.[i] Rome did not possess enough soldiery to prevent invaders from violating the periphery of its geographically extended empire. For centuries, however, it kept itself safe by the credible expectation that, in the event of an armed incursion, the Roman Legions would eventually respond and punish the transgressor. This was not a bluff on Rome's part. The empire didn't hesitate to launch punitive expeditions against tribes invading Rome's widely distributed borders as well as against rebellions within.

Here is some hair-raising advice offered two millennia later by Sir John Fisher, First Sea Lord, Admiral of the Fleet, and widely regarded as the most important British naval figure after Horatio Nelson. It emphasizes that deterrence by punishment is likely to be effective in proportion as the threatener has a fearsome reputation:

> If you rub it in both at home and abroad that you are ready for instant war . . . and intend to be first in and hit your enemy in the belly and kick him when he is down and boil your prisoners in oil (if you take any), and torture his women and children, then people will keep clear of you.[1]

In some cases, conventional deterrence operated by a combination of threatened punishment and active denial. This underpins classical "balance of power" strategy, whereby aggression is to be prevented by maintaining sufficient force that neither of two contending antagonists will calculate that aggression would be in its interest. For example, consider the centuries-old English policy of being the "balancer" by allying itselfwhich ied—counterintuitively, it appears at first glance—with the *weaker* side in continental Europe, thereby keeping any one country or alliance from becoming so dangerously powerful as to threaten the United Kingdom itself.

This strategy, interestingly, did not operate by the UK acting as a direct threat to a would-be belligerent, but by ensuring that the balance of power was sufficiently close that a kind of peace (or suspended antagonism) would prevail via the threat of otherwise prolonged and pointless warfare if anyone gets any dangerous ideas.

Exaggeration of foreign threats in the interest of justifying an intimidating but costly military posture is both historically deep and geographically wide. In his extended essay, "Imperialism and Social Classes," economist and historian Joseph Schumpeter described such excesses on the part of Rome's rulers, which should serve as a warning in our enemy-prone and threat-credulous twenty-first century:

---

[i] The word "deter" comes from the Latin "de" (away) as in detour, plus "terre" (fear), as in terror. To deter is to frighten someone or something away.

There was no corner of the known world where some interest was not alleged to be in danger or under actual attack. If the interests were not Roman, they were those of Rome's allies; and if Rome had no allies, then allies would be invented. When it was utterly impossible to contrive such an interest—why, then it was the national honor that had been insulted. . . . The whole world was pervaded by a host of enemies, and it was manifestly Rome's duty to guard against their indubitably aggressive designs.

One thousand five hundred years after the fall of the Roman Empire, General Douglas MacArthur, no peacenik he, observed, "Our government has kept us in a perpetual state of fear, kept us in a continual stampede of patriotic fervor—with a cry of a grave national emergency. Always there has been some terrible evil at home or some monstrous foreign power that was going to gobble us up."[2]

Although the United States is not unique when it comes to threat inflation, here is an abbreviated list of some events to be kept in mind as we ponder its universality:

- Manipulation of legitimate fear of terrorism into a "worldwide terrorist threat," particularly by the administrations of George W. Bush and Donald Trump, inflating it into a "war against terror" rather than a reason for international law enforcement.
- The George W. Bush Administration's trumped up claim that Saddam Hussein was involved in the attacks of 9/11 (and therefore constituted a threat of initiating more such outrages) and—even more threatening, and equally untrue—that Iraq possessed "weapons of mass destruction."
- Ronald Reagan's warning that the Sandinista government in Nicaragua constituted a threat to the entire Western hemisphere, including the United States, because it was "just two days' driving time from Harlingen Texas." A similar claim, already mentioned, was that the leftist government of Grenada threatened the safety of US medical students in that Caribbean country and was also constructing a runway from which Soviet bombers could attack the United States.
- Lyndon Johnson's dramatic broadcast to the nation in 1964, maintaining (falsely) that US naval vessels had been attacked in the Gulf of Tonkin, which served as a pretext for escalating the Vietnam War.
- Dwight Eisenhower's announcement in 1951 that "If Indochina falls, the countries of southeast Asia and Indochina will follow, followed by India . . ."
- Donald Trump's uppercase hyperventilating—by tweet—to the Iranian President on July 22, 2018, "NEVER, EVER THREATEN THE UNITED STATES AGAIN OR YOU WILL SUFFER CONSEQUENCES THE LIKES OF WHICH FEW THROUGHOUT HISTORY HAVE EVER SUFFERED BEFORE," combined with numerous prior warnings about how Iran threatened the entire Middle East, and,

because of the nuclear deal orchestrated by the Obama Administration, was about to obtain nuclear weapons.

- And, of course, Trump's saber-rattling "fire and fury" threats toward North Korea, preceded by numerous warnings that Kim Jong-un's government was on the verge of attacking the United States with nuclear weapons. This threat inflation subsequently morphed into excessive threat diminution when, after a brief meeting in Singapore with Mr. Kim during summer 2018, Trump tweeted that North Korea suddenly was "no longer a nuclear threat." In this case, what was inflated was Trump's self-presentation as a threat-diminishing alpha male, first posturing as defender against North Korea by exaggerating its danger to the American homeland, and then as one who has successfully removed that danger . . . when he hadn't.

As with the now-infamous statement by UK Prime Minister Neville Chamberlain, as he returned from his disastrous Munich Conference of 1938 with Adolf Hitler—when Chamberlain announced that he had achieved "peace for our time"—the pendulum has occasionally swung the other way, with international threats being under- instead of overplayed. This, however, has been exceedingly rare in world events. Far more often, threats have not only been the currency of international diplomacy, but often an opportunity to agitate a leader's domestic constituency, typically to justify military expenditures, enhance political support, and, not uncommonly, to set the stage for war.

A theme of this book has been that threats often end up making things worse, for threatener and threatened alike. To a degree only rarely known to the general public, international threats have precisely this history. Here is a predominantly nonnuclear example. In 2008, the United States was convinced that Russia, seeking to reestablish its international status, was planning to attack or at least politically intimidate its southern neighbor, the Republic of Georgia. The US accordingly advocated bringing Georgia into NATO, accelerated training of the Georgian military, and issued numerous warnings to the Putin government, believing that such threats would deter Moscow from any aggressive adventurism. The effect was precisely the opposite: the Russians were particularly and, in retrospect, predictably alarmed by the specter of an assertive Georgia joining NATO and were determined not to be intimidated by the United States. Meanwhile, Georgia was emboldened to intervene militarily in South Ossetia, a breakaway region with substantial pro-Russian sympathies that had sought to dissociate itself from the Tblisi government. Russia responded with overwhelming military force, not only ignoring the preceding US-issued threats, but in large part responding to them by doing precisely what Washington had sought to avoid.

A similar pattern occurred in Iraq as a result of an intentional but misguided policy that backfired. As professor Juan Cole observed,

ISIL did not arise organically from Islam. There was little violent religious extremism among Sunnis in Iraq before the United States invaded in 2003. Iraq had had a secular, socialist ideology and its government refused to put Islam in the constitution as the religion of state. Bush installed a Shiite sectarian government allied with Iran and pushed Iraq's Sunni Arabs into such despair that some of them turned to "al-Qaeda in Mesopotamia," which morphed into the "Islamic State of Iraq," and then after the Syrian revolution of 2011 became the "Islamic State of Iraq and the Levant." Had there been no American invasion and occupation of Iraq, there would have been no ISIL. US warmongering has sown dragons' teeth throughout the Middle East.[3]

When it comes to threats, the Law of Unforeseen Consequences could be amended to the Law of Self-Defeating Outcomes.

For another example, consider threatening a country—say, Iran—with severe economic penalties or even war in the hope of getting it to bend to the threatener's will. All too often, the result is to strengthen the hand of hardliners on the other side, leading to increased confrontation rather than accommodation. Later, we consider a related phenomenon: the "security dilemma," whereby a country—seeking to enhance its security—beefs up its military posture, which causes competing countries to feel threatened and that respond by their own military machinations. The outcome of this threat–counterthreat sequence is that everyone ends up more threatened and, thus, worse off.

The specter of "international communism" has repeatedly served as a bogeyman that bolstered US military spending, alliance-building, support of right-wing dictatorships (e.g., in Chile, Argentina, Brazil, Nicaragua, Iran, the Philippines, South Africa, Indonesia), plus the overthrow or attempted overthrow of left-leaning governments (e.g., Chile, Iran, Guatemala, Venezuela). In all of these cases, the consequences have been worse than if threats had not been perceived and if these perceptions had not led to self-defeating actions.

An honest assessment of the abuses and uses of threats also requires us to consider "The World According to Gap": the deceitful construction of so-called Bomber Gaps, Missile Gaps, Civil Defense Gaps, Throw-Weight Gaps, Windows of Vulnerability, and so on, caricatured in the movie *Dr. Strangelove* as, among other things, a "mine shaft gap" by which the Soviets were reputed to enjoy an advantage when it came to sequestering its population during a forthcoming World War III. In real life, these and other nonexistent gaps were used to escalate the nuclear arms race vis-à-vis the USSR during the Cold War.

With the end of the Cold War and the dissolution of the Soviet Union, contemporary US threat purveyors have been challenged to fill that particular gap, using by turns Cuba, Libya, Iraq, North Korea, international terrorism, fundamentalist Islamic extremism, Russia and China (redux), North Korea, Venezuela, and Iran. At the same time, notwithstanding the egregious tendency for political leaders to exaggerate threats—especially for domestic political gain—challenges to security can also be genuine. Or, as the sardonic saying has it, "Even paranoids have enemies." (Not to be forgotten, as well, is that their behavior often creates enemies.)

NUCLEAR DETERRENCE 101

"Strike through the mask!" was Captain Ahab's exhortation; pierce the superficial and get to the underlying reality of his nemesis. When it comes to international threat-making and the dangers they pose to threatener and threatened alike, nuclear weapons are—or should be—our Moby Dick. If you think that England's Bloody Code was too darned bloody, you ain't seen nothing yet. The updated nuclear version also relies on deterrence by punishment, except that its eighteenth-century English predecessor merely sought to punish each individual perpetrator, whereas deterrence by punishment in the nuclear sense would be a bit more consequential.

"You may not be interested in war," wrote Leon Trotsky, "but war is interested in you." This warning was chillingly appropriate in Trotsky's world of conventional weapons. In the current nuclear environment, where civilians are prone to assume that war isn't their business, it demands everyone's attention even more, whatever your level of interest.[ii] Notable among those especially interested in nuclear war are those who believe that the benefit provided by nuclear weapons exceeds the threat they pose, pointing to deterrence as a regrettable but ultimately worthwhile posture that repays whatever downside it presents.

Abundant psychological research has suggested interesting differences between conservative and liberal mindsets, notably that conservatives are generally more sensitive to threats—or, as liberals might say, more inclined to exaggerate them, and, as a result, to overrespond.[4] At the same time, conservatives are likely to claim that liberals are insufficiently responsive to such threats. In any event, this apparent divergence is consistent with the fact that, compared with liberals, conservatives are particularly liable to favor the death penalty, to oppose restrictions on gun ownership, to believe in hell as punishment for sinners, to be national populists and—especially relevant to the current discussion—to advocate nuclear deterrence. To some degree, of course, liberals (such as myself) can argue that we simply perceive different threats: those posed

---

[ii] On a backpacking trip in Alaska, I once encountered a fellow hiker who hadn't bothered to store his food safely. When I pointed out the risk he was taking, he replied that he wasn't interested in grizzly bears. I later learned that he had been severely mauled that night.

by use of the death penalty, by the existence of a widespread gun culture, by national populism, and, most especially, by nuclear deterrence itself.

This section seeks to strike through the mask of nuclear deterrence, showing how it poses unacceptable risks, in many cases actually *increasing* the likelihood of nuclear war. Nuclear deterrence is also extraordinary in the extent to which it has long been hidden behind its many masks, such that some serious unmasking is overdue. In his poem "Whispers of Immortality," T.S. Eliot struggled to see "the skull beneath the skin." To see the skull beneath the various masks that deterrence wears, you have to "get over yourself" and pay attention to things that most people would rather avoid. You will be challenged to overcome what has been called the "unthinkability bias"—a notion attributed to film critic Pauline Kael, who was allegedly surprised at the election of Richard Nixon, because she only knew one person who had voted for him! People are unconsciously biased against what they find unthinkably bad, to which must be added the illusion that if you don't think about something and, moreover, if no one you know is thinking about it, then it couldn't happen. In this sense, resistance to nuclear war faces a greater challenge than does opposition to global climate change, because at least the latter is happening *now*, abundantly confirmed by anybody not blinded by ideology. And yet, even though many of the costs—financial, environmental, even health related—are clear, resistance to this particular resistance movement continues, especially in the United States.

How much more difficult, then, to generate and sustain resistance to an even greater threat that hasn't yet occurred! Although dire science-based warnings about anthropogenic global climate change have been common currency since the mid 1980s, atmospheric carbon dioxide levels have continued to rise, as has the danger of a hot planet and rising seas. More or less simultaneously, nuclear weapons levels have gone down, from around seventy thousand to around sixteen thousand, although the danger of nuclear war—as estimated, for example, in the "Doomsday Clock" of the Bulletin of the Atomic Scientists—has, if anything, risen.

Numbers are key when it comes to global climate change because heating is a direct function of the accumulation of heat-trapping greenhouse gases. They are less important when it comes to nuclear war, because even a small number of detonations would produce worldwide catastrophe. Far more critical are those circumstances liable to precipitate such an event: the nature of the delivery and warning systems, the international political climate, and the underlying doctrine for their use. It is possible to plan ahead, thereby mitigating the effects of global climate change, even though it cannot be prevented (because it has already begun). Alas, there is no meaningful way to mitigate the effects of nuclear war, although it is possible to prevent it from happening in the first place. The most misleading and, in fact, counterproductive attempt at prevention has involved striving for deterrence by denial, a kind of nuclear age Great Wall of China, or Maginot Line.

In pursuit of strategic missile defense since Reagan's initial support of the program, the United States spent more than $330 billion as of 2017, by the estimate of Stephen I. Schwartz, a military analyst at the Middlebury Institute of International Studies in Monterey, California. Add to this, roughly $10 billion to $15 billion per year since then. Yet "Star Wars"—an astrodome from sea to shining sea—is still a pipedream. (Not surprisingly, the US government does not keep track of total expenditures on missile defense, or, if it does, the numbers have not been made available.)

After the US nuclear bombings of Hiroshima and Nagasaki in 1945, war changed. Until then, the overriding purpose of military forces had ostensibly been to win wars. "But now," wrote US strategist Bernard Brodie, "its chief purpose must be to avert them. It can have almost no other useful purpose."[5] Thus, nuclear deterrence (mostly abbreviated from here on as simply "deterrence") was born, a seemingly rational arrangement by which peace and stability were to arise by making the most apocalyptic threat of all: mutually assured destruction, appropriately abbreviated as MAD. Winston Churchill described it with characteristic vigor: "Safety will be the sturdy child of terror, and survival the twin brother of annihilation."

Robert Oppenheimer, the physicist who was overall scientific director of the Manhattan Project, likened mutually threatened nuclear annihilation to two scorpions in a bottle, each capable of killing the other, but only at risk of losing its own life in the process.[6] A chillingly negative image, but one that also relies on an unspoken and as yet untested assumption: that, under these conditions, each creature will restrain itself. One danger, although not the only one, is that the anticipated deterrence by punishment can result, not in mutual restraint, but in mutual punishment.

Deterrence is supposed to work by guaranteeing that any attacker will evoke retaliation so devastating that neither scorpion will attack in the first place. According to deterrence theorists, moreover, such mutual terror should be embraced, because insofar as it is shared, it would be mutually reinforced and thus maintained. Those who place their bets on Churchill's claim of safety and survival via deterrence cannot be proved wrong, because their ultimate wrongness wouldn't be subject to postmortem criticism, just as for the same reason, deterrence skeptics who warn about an apocalyptic future will not survive being proved right. And yet, the formal justification of the whole nuclear enterprise rests on this terribly shaky ground. The only way to repudiate an apocalyptic future is to repudiate nuclear weapons, and that requires repudiating deterrence.

So long as advocates and skeptics—nuclear Hawks and Doves—continue to occupy the same limbo, both agree that the core of deterrence is threat: to murder millions. Hence, nuclear deterrence is effectively a form of state-sponsored nuclear terrorism. As long-time US nuclear negotiator Fred Iklé pointed out, it relies on "a form of warfare universally condemned since the Dark Ages—the mass killing of hostages."[7]

Rarely appreciated, as well, is that during the course of maintaining such an over-whelmingly threatening posture, nuclear deterrence freezes existing hostilities among those countries ostensibly being deterred, thereby deterring—or, at minimum, greatly inhibiting—progress toward genuine reconciliation. Good fences may sometimes make good neighbors; nuclear deterrence does not.

John Quincy Adams, then Secretary of State, argued in 1821 that the United States was a "well-wisher to the freedom and independence of all," and that accordingly, "she goes not abroad in search of monsters to destroy." Since 1945, the hope of many, espe-cially in the strategic community, has been that the United States has created its own protective monster here at home—an aspiration that has thus far failed the tests of history and logic. A more suitable monster metaphor comes from the various tales of the Golem, a Frankenstein's monster type of creature made of clay for the purpose of defending different Jewish communities of medieval Europe. In the best-known of these myths, from Prague, the Golem eventually goes on a murderous rampage and must be forcibly deactivated.

A modern Golem, nuclear deterrence was born of fear, created to feed on threat: the threat that, if attacked, the victim will retaliate with great destruction, plus fear of an opponent who is dangerously armed, and whose intentions are malign and threaten-ing. The greater the fear and the more frightening the threat, the greater the tendency to cling to the nuclear Golem as a deterring counterthreat. And the greater the threat, the greater the fear. The result is the most vicious of all vicious circles, the outcome of which is that threats and appeals to fear have trumped what should be a fear that is far better grounded: fear of deterrence itself—that it will fail and, with it, much of life on Earth.

Despite the fact that deterrence remains an article of faith among the "realists" who have long orchestrated US strategic policy and who continue to do so, despite its inco-herence and instability, much of this faith is lip service only, analogous to deeply reli-gious individuals who profess belief in heaven, yet rarely rejoice when a loved one dies. Thus, if the US government really believed in nuclear deterrence—or in the billions of dollars spent on Ballistic Missile Defense—there wouldn't be such hyperventilat-ing about the threat posed by a nuclear armed North Korea or possibly by Iran in the future. Belief in nuclear deterrence is thus a mile wide but no more than an inch deep; its supporters are fair-weather friends, whose pessimism is, alas, well founded. The rest of this section seeks to demonstrate why.

Few things have been advertised more persistently, vigorously, and effectively than nuclear deterrence. And few things better reflect the triumph of fantasy over reality, whereas none has the potential of punishing us more severely. And yet, devotees of deterrence constantly reinforce each other's views in their own mutually constructed hall of mirrors, or intellectual echo chamber, insisting they are seeing deterrence in a uniquely clear-eyed and hard-headed way. "One thinks that one is tracing the outline of a thing's nature over and over again," wrote Ludwig Wittgenstein, "but one is merely tracing round the frame through which we look at it."[8]

Here is a basic way to look at the nature of our nuclear "thing": multiply something by a million and it hasn't only been changed quantitatively, but also qualitatively. Someone might currently have, say, forty dollars in her wallet; multiply it by ten and her newly acquired four hundred dollars would be impactful, but not likely to be life changing. Multiply by a million, however, and there's a good chance her future would be altered. Most people walk at about two miles per hour. Increase this by a factor of ten, and you are riding a bicycle at twenty miles per hour, or perhaps in a car going slowly. Multiply by a million, and you have exceeded escape velocity and are heading for outer space—a change, not just of degree, but of kind.

As Napoleon once pointed out, quantity has a quality all its own. (This, in the context of his bragging to the Austrian Empire's Foreign Minister, Klemens von Metternich, "You cannot stop me. I spend 30,000 lives a month.")

TNT is a powerful but conventional explosive; a ton can do enormous damage. Atomic bombs are measured in kilotons (thousands of tons of TNT), and hydrogen bombs in megatons (or millions of tons). The difference, once again, is qualitative, not merely quantitative. (By contrast, the so-called Mother of All Bombs—the largest conventional bomb, dropped by the United States on ISIS caves in Afghanistan in 2017—has a blast yield of eleven tons.) The temperature inside a nuclear explosion is similarly a thing unto itself—in the millions of degrees, something not otherwise found on Earth. Although deterrence exists in both conventional and nuclear incarnations, and even ignoring the latter's unique radioactive component, the quantity of potential nuclear destruction has a quality, à la Napoleon, all its own.

It might be that deterrence is indeed one way of obtaining security, just as arson might be one way of obtaining warmth.[iii] In the remaining pages, we shall unmask this travesty of logic and strategy, and, following Wittgenstein, look at the real nature of nuclear deterrence, rather than merely tracing 'round its frame.

Let's start.

A MULTITUDE OF MYTHS

Examining nuclear deterrence not only involves unmasking a clear and present danger, but also it provides entry to some rich and fascinating intellectual explorations, worth attention even if the fate of Earth wasn't literally in the balance. We begin by

---

[iii] Historian Robert Conquest once quipped, "Stalinism might be one way of attaining industrialization, just as cannibalism might be one way of attaining a high-protein diet."

considering the most widely asserted pillar of deterrence: that it has worked. Advocates insist that we can thank it for avoiding a third world war even when tensions between the United States and the USSR ran high. Then, as now, Western governments in particular were reluctant to acknowledge the salience of threat, preferring to call nuclear weapons and their role "deterrents" or "the deterrent," which encourages the public to define them as its proponents desire. But in fact, they are, above all, threats—to blast, burn, irradiate, incinerate, and, in many other ways, devastate civilian populations. Some enthusiasts even maintain that nuclear deterrence set the stage for the fall of the Soviet Union and the defeat of communism. In fact, one reason for the USSR's collapse appears to have been its desperate efforts to keep up with American nuclear weaponry. In this telling, the West's nuclear threats also prevented the USSR from invading Western Europe and delivered the world from communist tyranny.

In a remarkable case of secular religiosity, deterrence has been credited with "a miracle that eluded Christ: ushering in peace on earth *without* goodwill toward men."[9]

There are many credible reasons why the United States and the former Soviet Union avoided destroying each other, most notably because neither side wanted to go to war in the first place. Indeed, the two countries never fought each other throughout the twentieth century.[iv] Singling out nuclear weapons for the fact that the Cold War never became hot is like saying a junkyard car, without an engine or wheels, never sped off the lot because no one turned the key. There is no way to demonstrate logically that nuclear weapons kept the peace during the Cold War, or that they do so now. By contrast, as we shall see, there is plenty of historical evidence of narrowly avoided disaster by mistake, malfunction, or miscalculation.

As with the widely discredited assumption that capital punishment will deter murder, such a superficial connection is exploited by asserting that the threat of a devastating nuclear response will deter an attack. But as Admiral Eugene Carroll has pointed out, "It does not follow that war has been deterred solely by the nuclear threat. There are many, many other practical military, political and economic factors which weigh against superpower conflict far more effectively than the incredible abstraction of nuclear deterrence."[10] Thus, following the Second World War, many diplomatic, political, economic, and cultural institutions were put in place that prevented war and kept the peace. When asked why Europe has enjoyed seventy-five years of peace (except for Kosovo and Ukraine), nuclear weapons maven Joseph Cirincione pointed to

---

[iv] However, the United States did join other Western nations in sending a small and unavailing military force to Siberia in support of the counterrevolutionary White Russians shortly after the First World War.

NATO, the European Union, the International Monetary Fund, the World Bank, American super-power status, global economic prosperity, reduced disease, crime, and additional factors. In other words, isolating nuclear weapons as having kept the peace is not just superstition, it is cherry-picking facts to support a pre-determined conclusion.[11]

Its endless adjustments, in an effort to provide intellectual coherence, even as it provoked nuclear arms races and conceptual confusion, have gone under many names, including massive retaliation, mutually assured destruction, escalation dominance, escalation equivalence, escalation control, flexible response, launch on warning, launch under attack, intrawar deterrence, extended deterrence, and minimal deterrence. Such jargon is often embroidered with a veneer of arcane language and impenetrable acronyms, along with a superficially impressive overlay of mathematics.

It is tempting to judge the depth of a river by whether you can see the bottom: the more obscure the water, the deeper it seems to be. Strategic analysis similarly gives the impression of being profound, and yet, shallow waters can carry loads of mud, thereby seeming deep, whereas they are merely filled with sediment. What follows in the rest of this book may or may not be limpid, but its goal is to plumb the shallows of deterrence discourse.

Strategic theorists engaged in fervent attempts to "scientize" and thus legitimize nuclear deterrence after the Second World War, attempting to lend an inherently bogus notion an aura of intellectual respectability. But it is mostly ideology, not unlike postmodernism and closer to religious fundamentalism. It appeals to wishful thinking and often outright ignorance, combined with the fetishization of precision where none exists, and is in that sense comparable to what Alfred North Whitehead called the "fallacy of misplaced concreteness." We might call it the fallacy of misplaced mathematization. What follows is an equation-free zone; but, rest assured, because deterrence theory has no precision, nothing will be lost.

When it comes to the alleged successes, potential failures, and the nuts and bolts of nuclear deterrence, practical considerations of psychology, politics, and geography have consistently been far more important than the niceties of theory. Unlike, say, India and Pakistan—both of which have nuclear weapons and have also had several conventional wars—the two Cold War opponents didn't share a common border or have conflicting territorial claims. Perhaps peace prevailed between the two superpowers simply because they had no quarrel that justified such destruction, even conventionally fought. There is no evidence, for example, that the Soviet leadership ever contemplated trying to conquer Western Europe, much less that it was restrained by the West's nuclear arsenal. *Post facto* arguments may be the currency of pundits, but are impossible to prove and offer no solid ground for evaluating a counterfactual claim, conjecturing why something has *not* happened. (If a dog barks in the night, we might be able to say with confidence that it did so because someone walked by. If it does not bark, however, we may never know the reason for its silence.)

Enthusiasts of the success of those threats labeled "nuclear deterrence" resemble the man who sprayed perfume on his lawn every morning. When a perplexed neighbor asked about this strange behavior, he replied, "I do it to keep the elephants away." The neighbor protested that there weren't any elephants within ten thousand miles, whereupon the perfume-sprayer announced, "You see, it works!" We looked earlier at the special attraction of religion as a salve in hard and threatening times (especially for those who survived, regardless of why they actually did so); attributing continued national survival to the successful influence of nuclear deterrence can be similarly reassuring, regardless of what actually kept the elephants at bay.

In 1948, psychologist B.F. Skinner conducted an experiment in which hungry pigeons were fed on a random schedule. Soon, three quarters of them were behaving in unusual ways, depending on what each had been doing just before getting food: one rotated her body (always counterclockwise), another swung his head like a pendulum, a third wiggled her feet, and so on. The resulting research report was titled "Superstition in the Pigeon."[12]

We and the Soviets threatened each other with mutual destruction if either attacked the other, and no war occurred between us—a seemingly impressive correlation. But correlations can be spurious, as with ice cream consumption and drowning. They are strongly correlated, but not because eating ice cream makes people drown, but because both events tend to happen in hot weather. If a pigeon spun around and didn't get fed, it would presumably have been disappointed, but no great harm would have been done, unlike the outcome if nuclear threats fail to achieve their end. It is difficult to give up a superstition. Perhaps something bad really will happen if we walk under a ladder, and perhaps we've been saved from disaster because we've avoided taking such a stroll.

Realism suggests otherwise. In ancient China, it was widely believed that solar eclipses were caused by a dragon swallowing the sun, so people responded to sudden darkening by scaring away the dragon by making as much noise as possible—banging pots and gongs, yelling loudly—and guess what? It worked! Every time. If, for some reason, people had refrained from all that noise-making and the eclipse resolved anyhow, the worst outcome would have been a loss of respect for the efficacy of dragons. But, if nuclear deterrence fails, its advocates will not be around to bemoan its ineffectiveness.

In some cases, it only takes one failure for an entire scaffolding, previously thought safe, to come crashing down. The Concorde supersonic transport entered service in 1976 and flew flawlessly throughout the late 1970s and 1980s. In fact, it was lauded as not only the fastest, but also the *safest* passenger plane of all, having a zero accident and fatality rate. Then, in 1990, one of them crashed on a runway in Paris, killing all 109 people on board and ultimately grounding the entire fleet, which was subsequently abandoned. Its safety record instantly jumped from the best to the worst (because only a handful of the planes were ever built and flown). The disaster occurred after a tire blew out, which initiated a chain of multisystem failures whereby the fuel tanks ruptured, from which there was no return. Interestingly—and tragically—this potential weakness in the Concorde system had earlier been identified, but nothing was done to correct it.[v] We'll look later at weaknesses in the threat system known as deterrence that have also been identified and that also remain unaddressed.

On the other hand, it has been pointed out—correctly—that the United States and Russia did not go to war with each other after nuclear weapons arrived on the international scene; but then, we didn't go to war with each other before that either. There have also been some seeming successes of deterrence, whereby it is widely believed that it functioned as hoped, that the threat of awful consequences has prevented war, not only nuclear but also conventional.

Take the Cuban Missile Crisis of 1962—when by most accounts we were closest to nuclear Armageddon—often cited as nuclear deterrence at its best. But, in fact, this crisis was *caused* by nuclear weapons, specifically the Soviet attempt to base nuclear-armed missiles in Cuba, because the United States had placed nuclear-armed Jupiter missiles in Turkey, targeting Moscow. According to many historians, the main reason Khrushchev eventually backed down was because the US quietly agreed to withdraw those missiles, and also because Soviet conventional military forces in the Caribbean were tiny compared to that of the United States. The USSR was also greatly outmatched in the nuclear arena at the time, being on the short end of the supposed missile and bomber gap that John F. Kennedy had grossly misrepresented to the American public during his 1960 election campaign.[vi]

The reality is that the Cuban Missile Crisis was resolved short of nuclear war not because of nuclear deterrence, but despite it. It may seem a truism that absent nuclear weapons there wouldn't have been any crisis, but that is precisely the point: Khrushchev's move to install nuclear weapons in Cuba in 1962 was a direct consequence of the Soviet perception that such weaponry was needed. Why this perceived need? To deter the United States, in two respects. For one, the failed Bay of Pigs invasion (sponsored by the United States) raised fear in the Kremlin—and Havana—that there could well be another attempted overthrow of Fidel Castro's communist government; Soviet

---

[v] Many thanks to Dr. Martin Hellman for pointing out the example of the Concorde.

[vi] Part of Kennedy's electoral success derived from his claim that Soviet missiles dangerously outnumbered those of the United States. This supposed "missile gap" was truly immense, but reversed: the US had hundreds of long-range missiles, and the first photoreconnaissance satellite (called *Discoverer*) shortly revealed that the Soviets had a grand total of four.

nuclear weapons would, it was hoped, deter any repeat adventurism. And second, the USSR was intensely aware—although most Americans were not—that prior to those Jupiter missiles in Turkey, the United States had deployed intermediate-range nuclear-equipped Thor missiles in the United Kingdom in 1959. Under the unspoken rules of deterrence, this perceived threat demanded a comparable response. And why had the United States placed those missiles in Europe? To deter the Soviet Union! An initially unpublicized part of the agreement that ended the Cuban Missile Crisis was for these Soviet and US missiles—i.e., mutual provocations—to be removed.

Although nuclear deterrence has yet to fail, history, alas, is not reassuring when it comes to anything approaching a triumphant trajectory. Foremost is the fact that the danger of annihilation wouldn't exist if not for those bombs and warheads, the existence of which has largely been mandated by deterrence. Moreover, there is no basis for confidence that the absence of nuclear war *thus far* presages its continued absence. Even if we assume that it has prevented global war (and as we'll see, there is little reason to make this assumption), how long can we reasonably expect that this will continue?

Bertrand Russell pointed out that one might credibly expect a tightrope walker to remain safely aloft for a minute, or ten minutes, or perhaps even longer. But for a hundred years?

Martin Hellman, Stanford University Professor, winner of the Turing Prize for Computer Science Innovation (that field's equivalent of a Nobel Prize), and Adjunct Senior Fellow for Nuclear Risk Analysis at the Federation of American Scientists, described nuclear risks this way, making the optimistic assumption of five hundred years:

> Even if nuclear deterrence could be expected to work for 500 years before we destroy ourselves, one-sixth of 500 years is 83 years. Thus, a child born today in America would have roughly one chance in six of being killed by a nuclear weapon over their expected lifetime—the same as in Russian roulette. Are we pointing a partially loaded revolver at the heads of the next generation of Americans? Is the time frame closer to 100 years? If so, are we spinning the cylinder and pulling the trigger five times during that child's expected lifetime?[13]

The so-called "nuclear peace" has been very brief so far; the interval between the Second World War and the end of the Cold War covers about five decades. More than twenty years separated the First and Second World Wars; before that, there had been more than forty years of relative peace between the end of the Franco-Prussian War (1871) and World War I (1914). Fifty-six years elapsed between Napoleon's defeat at Waterloo (1815) and Bismarck's war with France. Or, taking a broader look at the historical record when it comes to big, pan-European wars, there had been a gap of nearly one hundred years from the end of the Napoleonic Wars (1815) to the beginning of the First World War (1914). Even in war-prone Europe, decades of peace have not been so rare—yet they always concluded with war, during which the combatants used those weapons that were available. The only exception was chemical weapons during the Second World War, in large part because they had been stigmatized and outlawed by the 1925 Geneva Gas Protocol, following widespread revulsion from their effects in the First World War.

The fact that chemical weapons, having been used in World War I, were not used in World War II might also be the result of a degree of deterrence. Italy, for example, didn't hesitate to use chemical weapons against the Ethiopians during the 1930s, just as Syria's Assad used them against antigovernment forces. On the other hand, chemical weapons have also turned out to be much less militarily useful than its early advocates had assumed, mostly because their effectiveness depends greatly on uncontrollable weather conditions—notably, wind speed and direction.

Writing of the new Maxim machine guns, *The New York Times* congratulated the Western nations editorially in 1897 for possessing such wondrous devices that are "peace-producing and peace-retaining terrors."[14] Peace-producing? No way. A year later, British writer Hilaire Belloc noted sarcastically, "Whatever happens we have got/ the Maxim gun and they have not." Shortly afterward, "they" (Germany and its allies) obtained Maxim guns and even more fearful weapons, none of which produced or retained peace when the First World War rolled around.

ALTHOUGH DETERRENCE HASN'T FAILED, IT HASN'T "SUCCEEDED" EITHER

Belief in the deterrent value of ferocious weapons is widespread and has been around for a long time. For example, Alfred Nobel, inventor of dynamite (which earned him a fortune), believed that its conveniently packaged and transported explosive power would make war obsolete because it would be so devastating as to be unthinkable. But when a French newspaper erroneously published an obituary, criticizing him for having "become rich by finding ways to kill more people faster than ever before," Nobel became concerned about his legacy and established the Nobel Prizes in an attempt to atone. Nuclear weapons are substantially more terrifying than dynamite.

Even when possessed by just one side, nuclear threats have not deterred war. The Chinese, Cuban, Iranian, and Nicaraguan revolutions all took place even though a nuclear-armed United States backed the governments that were eventually overthrown. Similarly, the US lost the Vietnam War, just as the Soviet Union lost in Afghanistan, despite the fact that both not only possessed nuclear weapons, but also more and better conventional arms than their adversaries. Nor did the implicit threat posed by its nuclear weapons aid Russia in its initially unsuccessful war against Chechen rebels in 1994 to 1996, or later from 1999 to 2000, when Russia's conventional forces devastated the suffering Chechen Republic. It was a nuclear-armed United States that "lost" China in 1949 and a nuclear-armed Soviet Union that lost China again during the late 1950s and early 1960s. (It is odd, incidentally, that such a huge nation can have been so frequently misplaced.)

Nuclear weapons have not enabled the United States to achieve its goals in Iraq or Afghanistan, which have become costly failures for the country with the world's most advanced strategic forces. Moreover, despite its nuclear arsenal, the US remains fearful of domestic terrorist attacks, which are more likely to be made *with* nuclear weapons than deterred by them. If, as some have suggested, terrorism is the nuclear weapon of the poor and powerless, nuclear weapons (and their supporting ideology of deterrence) are the terrorism of the rich and powerful. At the same time, the appeal of an alpha leader, so important to national populists, is paralleled by the appeal of nuclear weapons for those who similarly find themselves struggling with a frustrating lack of reassuring identity and for leaders who worry that they and their country have been losing their legitimate place in the sun.

We have already seen that, in some cases, the threat of eternal damnation in the hereafter has motivated some pious personal behavior. But we cannot conclude similarly that nuclear weapons have deterred *any* sort of war, and although "past performance is no guarantee of future returns," they probably will continue not doing so in the future. During the Cold War, each side engaged in conventional warfare: the Soviets, for example, in Hungary (1956), Czechoslovakia (1968), and Afghanistan (1979–1988); the Russians in Chechnya (1996–2009), Georgia (2008), Ukraine (2014–present), as well as in Syria (2015–present); and the United States in Korea (1950–1953), Vietnam (1962–1974), Beirut (1982), Grenada (1983), Panama (1989), the Persian Gulf (1990–1991), the former Yugoslavia (1999), Afghanistan (2001–present), and Iraq (2003–2016), to mention just some of the more prominent cases.

In 1948, the USSR restricted access to West Berlin, which was jointly administered at the time by the United Kingdom, France, and the United States. Also at that time, the US had a nuclear monopoly; this didn't inhibit Stalin from initiating the Berlin Blockade, one of the most provocative and aggressive actions of the Cold War. In fact, the USSR was most aggressive vis-à-vis the United States between 1945 and 1949, when only the United States had nuclear weapons. It was during that time that Stalin, in violation of the promises he had made to Roosevelt and Churchill during their Yalta summit, consolidated Soviet control over its Eastern European satellites. All this while the United States "enjoyed" a worldwide nuclear monopoly.

Nor has the threat of another side's atomic and then hydrogen arsenals deterred actual attacks by nonnuclear opponents upon nuclear armed states or their avowed strategic interests. In 1950, China was fourteen years from developing its own nuclear weapons, whereas the United States had dozens, perhaps hundreds, of atomic bombs. US military and civilian officials judged, moreover, that China's military was exhausted by decades of civil war and would not dare intervene against the world's sole nuclear superpower. They were spectacularly wrong. As the Korean War's tide shifted against the North, Mao's China felt threatened that General MacArthur's forces wouldn't stop at the Yalu River and would invade China in an attempt to overthrow its communist government. To the surprise and consternation of the US leadership, the American nuclear arsenal did not deter China from sending more than three hundred thousand soldiers southward, resulting in the stalemate on the Korean peninsula that divides it to this day, and that has produced one of the world's most dangerous, unresolved standoffs.

In 1956, nuclear-armed Great Britain warned nonnuclear Egypt to refrain from nationalizing the Suez Canal—to no avail. The United Kingdom, France, and Israel ended up invading the Sinai in an unsuccessful effort to achieve their goal. Argentina attacked the British-held Falkland Islands in 1982, even though the United Kingdom had nuclear weapons and Argentina did not. Following the US-led invasion in 1991, conventionally armed Iraq was not deterred from lobbing thirty-nine Scud missiles at nuclear-armed Israel, which did not retaliate, although it could have vaporized Baghdad. It is hard to imagine how doing so would have benefited anyone. The fact that Israel had this capacity did not stay Saddam's hand, perhaps because he realized that Israel would have had more to lose than to gain by "making good" on its unstated but universally understood deterrent threat. Moreover, nuclear weapons obviously did not deter the terrorist attacks of 9/11 on New York and Washington, DC, just as the nuclear arsenals of the United Kingdom and France have not prevented repeated terrorist attacks on those countries.

The pattern of nuclear nondeterrence is historically established and geographically widespread. Nuclear armed France couldn't prevail over the nonnuclear Algerian National Liberation Front. The US nuclear arsenal didn't inhibit North Korea from seizing an American intelligence-gathering vessel, the USS Pueblo, in 1968. Even today, this ship remains in North Korean hands. Its nuclear arsenal didn't enable China to get Vietnam to end its invasion of Cambodia in 1979; a conventional invasion did. Nor did US nuclear weapons stop Iranian revolutionary guards from capturing US diplomats and holding them as hostages from 1979 until 1981, just as fear of American nuclear weapons didn't prevent Iraq from invading Kuwait in 1990.

Moreover, the historical record is clear that when a nuclear state is losing in an armed struggle against a nonnuclear one, being armed with what was once called "the winning weapon" doesn't contribute to winning. The United States unequivocally lost in Vietnam, but accepted this defeat rather than flailing about with its atomic and hydrogen bombs. Ditto for the USSR in Afghanistan. And, although no one would be surprised if the United States "loses" Afghanistan as well, no one expects this outcome will be reversed by incinerating Kabul.

By the end of the twentieth century, both India and Pakistan had nuclear weapons, which might have inhibited each side—thus far—from using them. But it certainly hasn't made their confrontations less dangerous, nor, it seems likely, any less frequent. In 1999, Pakistan snuck military units—disguised as Kashmiri militants—into an area known as Kargil, a high-altitude region on the Indian side of the Line of Control that separates India and Pakistan in the disputed region of Jammu and Kashmir. The Pakistanis apparently thought that their nuclear arsenal would force India to accept the move as a *fait accompli*. Pakistan had tested its first nuclear weapons the previous year, and it seems likely that its military was emboldened by this addition to its arsenal, expecting that the threat of going nuclear would inhibit an Indian response. If so, it didn't work. India responded by mobilizing two hundred thousand troops, initiating an air campaign (not answered by Pakistan) and preparing a naval blockade of Karachi.

Pakistan's next step was to begin issuing nuclear threats. Prime Minister Nawaz Sharif announced, "If there is a war, or if the present confrontation continues on the borders, it will bring so much devastation, the damage of which will never be repaired." This did no good whatsoever and, by mid June, Indian forces had retaken all of the key positions in Kargil. India's nuclear arsenal had not deterred the Pakistanis from their military adventuring, just as Pakistan's didn't prevent India from retaking its lost territory.

Earlier, we briefly reviewed the compelling evidence that the threat of capital punishment has not deterred murder. It seems equally apparent that the threat posed by nuclear deterrence has not deterred war (although, admittedly, we only know about wars that have happened, not those that might have been prevented). What about the possible effectiveness of nuclear threats in enabling countries to get their way?

The United States didn't initially develop nuclear weapons for deterrence (a doctrine that didn't formally exist at the time), but to make sure that Nazi Germany didn't do so first, and then, it is claimed, after Germany was defeated, to end the Second World War by coercing Japan to surrender. For some historians, however, the bombing of Hiroshima and Nagasaki wasn't so much the conclusion of that war but the beginning of the Cold one: to impress Stalin, thereby preventing the USSR from pushing its postwar manpower advantage in Europe.

We come, therefore, to the question of nuclear coercion. Although deterrence seeks to prevent a country from doing something that it might otherwise attempt, the goal of coercion is to twist a country's arm, getting it to do something that otherwise it would not. (Shades of using hell to get people not only to refrain from sinning, but to do things they might not otherwise.) Threats could, in theory, serve as a useful manipulative tool in a country's diplomatic arsenal. At the same time, leaders worry about being coerced, manipulated, or deterred by their opponents. American strategists are concerned, for example, that new nuclear states, such as North Korea, and perhaps eventually Iran, might be able to exert their own coercion against the United States—that even a fledgling nuclear weapons state might feel empowered to bully or invade its neighbors, while at the same time its own nuclear forces, even if relatively small, would keep the US from interceding.

Nuclear coercion turns out to be more difficult, however, than it might seem, and not just because of the problem of credibility (to be explored shortly). This is because when one side is attempting to coerce the other, the goal is to get the recipient to modify the status quo, to *do* something. By contrast, deterrence aims to *prevent* the other side from doing something that would modify the status quo, and it is generally easier to refrain from an act than to carry one out. Moreover, pretty much by definition, a would-be coercer has been living with the current situation, which diminishes the credibility that it might suddenly initiate nuclear war to change things.

In *Nuclear Weapons and Coercive Diplomacy*,[15] political scientists Todd Sechser and Matthew Fuhrmann examined 348 territorial disputes that occurred between 1919 and 1995,[vii] using statistical analysis to see whether nuclear-armed states were more successful than conventional countries in coercing their adversaries during territorial disputes. They weren't. Not only that, but nuclear weapons didn't embolden those who own them to escalate demands; if anything, such countries were somewhat *less* successful in gaining their ends. In some cases, the researchers' efforts to make their analysis complete and unbiased becomes almost comical. For example, among the very few cases in which threats from a nuclear-armed country were coded as having compelled an opponent was the United States insistence, in 1961, that the Dominican Republic hold democratic elections after the assassination of dictator Rafael Trujillo. Another successful demand, this time in 1994, followed a military coup in Haiti, whereupon the Clinton Administration insisted that the rebellious colonels restore President Jean-Bertrand Aristide to power. Similarly, in 1974 to 1975, nuclear China forced non-nuclear Portugal to surrender its claim to Macau. These cases were included because professors Sechser and Fuhrmann sought, honestly, to consider all instances in which a nuclear-armed country got its way vis-à-vis a nonnuclear one. But no serious observer would attribute the capitulation of Portugal or of the Dominican Republic to the nuclear weapons possessed by China and the United States, respectively.

[vii] The prenuclear disputes provide useful background data on the proportion of nonnuclear threats that have succeeded in the past.

Here are Sechser and Fuhrmann's basic conclusions[viii]:

1. Nuclear states are not more likely than nonnuclear states to escalate their demands during a crisis; that is, possessing nuclear weapons does not provide states with leverage not otherwise available. Nuclear powers are likely to escalate their demands in the midst of a territorial crisis nineteen percent of the time. Nonnuclear states? Twenty-four percent. Even after using conventional force, nuclear states are no more successful than are their nonnuclear counterparts. The United States seriously considered trying to use atomic bombs to help raise the siege of French forces at Dienbienphu, the battle that ended France's colonial domination of Vietnam, but thought better of it.

2. Nuclear states are no more likely to succeed in whatever demands they do make than are nonnuclear states. Conversely, nonnuclear states are not more likely to make concessions when their adversary is a nuclear state. For example, China developed nuclear weapons in 1964; between 1949 and 1963, China gave substantial ground in seventy percent of its territorial disputes, compared with a seventy-seven-percent rate of backing down between 1964 and 2005. Sechser and Fuhrmann note that "these figures actually suggest that China's bargaining power lessened somewhat after it built a nuclear arsenal." On the other hand, it is at least possible that Russia's nuclear weapons emboldened Putin to annex Crimea, although explicit nuclear threats were not issued.

3. Their data set enabled Sechser and Fuhrmann to assess whether states that lacked nuclear weapons earlier during the twentieth century experienced greater success after they developed a nuclear arsenal. There was no such effect. Although a threat of conventional war is relatively easy to issue, and is often taken seriously, it is very difficult for a country to engage in credible nuclear blackmail. In 1958 and into 1959, Soviet Premier Khrushchev sought to end the West's rights to station troops in West Berlin, essentially repeating Stalin's unsuccessful gambit of ten years earlier, only this time, the Soviets were nuclear armed. According to a 1960 CIA report, Khrushchev told Averill Harriman, US Ambassador to the USSR, that "the West seemed to forget that a few Russian missiles could destroy all of Europe. One bomb was sufficient for Bonn and three to five would knock out France, England, Spain and Italy." He continued, "We will put an end to your rights in Berlin and our rockets will fly automatically"[16] if the United States resists Soviet demands. But the US did resist, refusing to knuckle under to Khrushchev, whose attempted coercion failed. It failed again in 1961, when United States and Soviet tanks were literally muzzle to muzzle during the second Berlin crisis.

[viii] The following is my synopsis, not their words.

It is often difficult for would-be coercers to send a clear signal. For example, during the 1999 Kargil War between Pakistan and India, the Pakistanis moved nuclear-capable missiles out of storage to potential launch sites. Was this an effort to coerce India by indicating that Pakistan might strike first if its demands were not met? Or was it an effort at deterrence, signaling that Pakistan would retaliate if its homeland were directly attacked? Or simply an attempt to reduce Pakistan's vulnerability if India struck first? If a message was in fact intended, what was it? Similarly, three decades earlier, President Nixon ordered an airborne alert of US strategic bombers, part of an effort to coerce Hanoi and Moscow to end the Vietnam War on American terms, but the intended recipients didn't realize this was even supposed to be a threat. On the other hand, it has been claimed (but not confirmed) that when Soviet forces refused to leave northern Iran after World War II, President Truman successfully used an atomic threat in 1946 to force Stalin's military to withdraw.

What about the coercive effect of the atomic bombing of Hiroshima and Nagasaki? According to the official US Strategic Bombing Survey, a formal analysis conducted immediately after the Second World War, Japan was on the verge of surrendering anyhow, and the atomic bombs did little to hasten that outcome. It has also been cogently argued that the real reason Japan surrendered when it did was not the atomic bombings, but because the USSR declared war on Japan on August 9, 1945, consistent with an earlier agreement between Stalin and Roosevelt.[17] For Japan, attributing its defeat to a powerful and hitherto unknown weapon was a convenient way to save face, and in this case, US interests coincided with those of the Japanese leadership, because giving war-ending credit to the atomic bomb burnished the image of US power, contributing to the impression that we possessed a mysterious "winning weapon."

But rather than giving up in the face of such a threat, other countries responded by producing their own. The USSR was spurred to develop its nuclear forces to deter the United States. The United Kingdom and France scrambled to join in, as did China, whose falling out with the Soviets had become intense by the late 1950s. India, having lost a brief war with China in 1962, developed a nuclear arsenal to deter its more powerful neighbor, whereas Pakistan went nuclear to deter India. North Korea's nuclear arsenal has been an effort to deter the United States, and if South Korea and/or Japan do the same, it will be to deter North Korea. Meanwhile, Israel developed nuclear weapons to deter the Arab states; if and when the latter go nuclear as well, no one will doubt why. Nuclear deterrence has thus been highly successful . . . as a driver of nuclear proliferation.

Well then, aren't nuclear weapons good for *something*? What about deterring terrorists? But how can people who do not fear violent death be deterred? And even assuming that retaliation would do a victim country any good, how can perpetrators lacking a fixed location even be targeted? Deterrence theory is therefore helpless when it comes to committed terrorists, many of whom are likely to be embedded in the very societies they are planning to attack, while their leaders are in remote locations, and/ or would probably not be deterred in any event because they do not fear—and may even welcome—martyrdom. How can people who believe their violence is sacred and will be rewarded either by God or by their extended community be deterred? Or even found and targeted for nuclear retaliation? A nuclear explosion in a crowded US city would be fiendishly difficult to trace because it would not include a return address or US Postal Service bar code, and, moreover, any potential evidence would almost surely be annihilated along with the victims.

Sidney Drell, renowned particle physicist, Deputy Director of the Stanford Linear Accelerator, and long-time senior government adviser on nuclear technology and weapons, noted in 2012, "In dealing with terrorists or rogue governments, nuclear deterrence doesn't mean anything—its value has gone. Yet the danger of the material getting into evil hands has gone up. So, what are existing nuclear arms deterring now? In this era, I argue that nuclear weapons are irrelevant as deterrence."[18]

Also problematic is that if nuclear weapons were used in response to a terrorist attack, the cycle of terrorist/counterterrorist violence would almost certainly spiral out of control, as tit-for-tat inspires an eye for an eye, a city for a city, a nation for a nation. In short, keeping terrorists at bay is simply not a job for deterrence. It should go without saying—but is worth saying nonetheless—that US nuclear weapons did not deter the terrorist attacks of 9/11, just as the nuclear arsenals of the United Kingdom and France have not prevented terrorist attacks on those countries.

CLOSE CALLS

We have looked at the claim that deterrence "works," examining the many holes in this assertion. And yet, hasn't it succeeded in the straightforward sense that, after all, we are still here? There has not been a nuclear war. By the same token, those of you reading this have not died, but would be ill-advised to take this as evidence you won't do so, some day. Nor is this logic limited to personal life and death. "Things that have never happened before," observes historian Scott Sagan, "happen all the time in history."[19]

Maybe defenders of deterrence are correct and it will never fail. There, doubtless, are people of goodwill who believe this, or yearn to do so. In his comic play *An Ideal Husband*, Oscar Wilde had one of his characters observe, "When the gods wish to punish us, they answer our prayers." Historian Daniel Boorstin wrote that "The American citizen . . . lives in a world where fantasy is more real than reality," before modifying Wilde's dictum to "When the gods wish to punish us, they make us believe our own advertising."[20]

Such believers only have to be wrong once; there are no mulligans in this particular game. "Defenders of nuclear deterrence need to prove they are right every time" writes Michael Krepon. He continues,

> Opponents of the excessive requirements undergirding nuclear deterrence only have to be proven right one time—with the survivors mourning the consequences. Given these odds, which side offers the most prudential choices? It's the side that loses more often than it wins. Fear of the Bomb has resulted in many unwise choices because those who place their faith in "strengthening" nuclear deterrence have yet to be proven demonstrably and disastrously wrong.[21]

We have come close many times, far more often and much closer than the public realizes. Here are just a few cases. The common pattern underlying all of them is that participants in mutual nuclear threats have been pushed by shortened time horizons to take the prospect of being attacked more seriously than the possibility that their own threatened retaliation will suffice to keep them safe.

Best known is the Cuban Missile Crisis of 1962, which we just visited in its mislabeled guise as a case of successful nuclear deterrence. Here is a bit more detail, revealing how close to disaster this crisis became and how little we can credit deterrence for the fact that nuclear war didn't break out. Soviet Premier Khrushchev sought to implant medium-range nuclear missiles in Cuba, acquiescing in part to a request from Fidel Castro, who, following the unsuccessful Bay of Pigs invasion sponsored by the Kennedy Administration in 1961, worried that another such attempt to overthrow his regime was in the offing. Castro hoped that the threat of Soviet nuclear weapons just offshore would deter the United States. In addition, the USSR was eager to match the prior US deployment of medium-range nuclear missiles close to its borders in Turkey and the United Kingdom. JFK demanded that the initial Soviet deployment be removed, and ordered a naval "quarantine"[ix] of Cuba. The world held its breath as Soviet naval vessels approached those of the United States, and exhaled with collective relief when the USSR turned around.

In the end, Soviet nuclear weapons and missile emplacements were removed, along with bombers, and the United States pledged not to invade Cuba. Secretly, US missiles were later removed from Turkey (something that had already been planned *before* the crisis, but Kennedy's advisers felt could not be done as a quid pro quo at the time, because it might be perceived as a sign of American weakness). In addition, a

---

[ix] The word "quarantine" was chosen because a "blockade" is considered an act of war.

"hotline" was established to minimize miscommunication between the United States and the USSR.

Prior to this resolution, things had been touch and go. The readiness level of US strategic forces was raised to DEFCON 2 (one notch short of war)—and for the first time ever, before or since, one eighth of the B-52 bomber fleet was placed on continuous airborne alert while B-47 bombers, also fully loaded with nuclear bombs, were dispatched to various military and civilian airfields. High-level negotiations between the two countries proved to be immensely stressful and confusing, and although the American public widely believes the Soviets simply backed down, the truth was much more complicated, with several events bringing the two countries even closer to nuclear war than was reported at the time. Here is a brief summary.

As the crisis intensified, a U2 spy plane was shot down by a Soviet-made anti-aircraft missile fired from Cuba. US officials believed this had been ordered by the Kremlin, leading some to conclude that war was imminent. It was subsequently revealed that the order had been given by Cuban officers at the scene, without knowledge or approval by the Soviets. As things heated up, important members of JFK's military and civilian structure—notably, General Curtis LeMay, Air Force Chief of Staff—began arguing for an immediate preemptive strike against the Soviet missile sites in Cuba. President Kennedy called in an experienced "wise man," President Truman's former Secretary of State Dean Acheson. According to the book, *Thirteen Days*, by Robert Kennedy (President Kennedy's brother, Attorney General at the time, and JFK's most important and trusted adviser), the following conversation ensued:

PRESIDENT KENNEDY: Dean, how does this all play out?

DEAN ACHESON: Your first step, sir, will be to demand that the Soviets withdraw the missiles within 12 to 24 hours. They will refuse. When they do you will order the strikes, followed by the invasion. They will resist and be overrun. They will retaliate against another target somewhere else in the world, most likely Berlin. We will honor our treaty commitments and resist them there, defeating them per our plans."

PRESIDENT KENNEDY: Those plans call for the use of nuclear weapons. So what is the next step?

DEAN ACHESON: Hopefully cooler heads will prevail before we reach the next step.

Unknown to US authorities, the Soviets had already put at least twenty nuclear-tipped missiles in Cuba,[22] capable of reaching as far as Washington, DC, each carrying a one-megaton nuclear warhead—equivalent to roughly seventy Hiroshima-size bombs. Had the president followed the advice of his military leaders, a nuclear war would almost certainly have resulted. In this regard, the failure of the Bay of Pigs invasion a year before (in 1961), may have been a godsend, because this disastrous enterprise—confidently urged by the CIA and the Joint Chiefs—soured President Kennedy on the reliability of advice by those whom JFK dismissively called the "brass hats."

At the same time, a Soviet submarine submerged off the Cuban coast was being harassed by a US Navy surface flotilla, which was firing small depth charges of the sort used in training exercises—not trying to destroy the submarine, but rather to get it to surface. The submarine's officers, however, believed they were under full attack; the US military was unaware that this sub was equipped with at least one nuclear-tipped torpedo. Unable to communicate with its military leadership in the Kremlin, the officers in the Soviet submarine had previously been given permission to use its nuclear weapon in "dire circumstances."

Accordingly, the captain, Vassili Savitsky, considered putting the ship's 15-kiloton nuclear torpedo in operational condition. He told his crew, "We're going to blast them now! We will die, but we will sink them all. We will not disgrace our Navy."[23] Two out of the three officers with launch responsibility voted to fire their nuclear torpedo at the US fleet, believing a war had already started. This is yet another thing that would easily have provoked a thermonuclear war. But the third officer, one Vasily Arkhipov, voted no, and so the Soviet submarine didn't devastate a chunk of the US Navy, the United States did not retaliate, and the world remained intact.

Forbearance by the Soviet undersea threat was, coincidentally, followed by a challenge to the Soviet "Air Defense Forces" by a US airborne threat. That same day, another U2 spy plane accidentally went off course and overflew Soviet Siberia. As a result, MIG fighter jets were scrambled, leading the United States, in turn, to launch F-102 interceptors armed with nuclear air-to-air missiles. By good fortune, the U2 got out of Soviet airspace before this incident escalated. When told of the events, JFK grumbled, "There's always some sonofabitch that doesn't get the word."

Anyone who hasn't already done so might want to see the movies *Fail Safe* and *Dr. Strangelove*, both of which depict nuclear disasters when, in fact, some sonofabitch doesn't get the word. Daniel Ellsberg, who was working at the RAND Corporation at the time, designing top-secret plans for carrying out a nuclear war, reported being shaken after seeing *Dr. Strangelove*, which he felt seemed less like fiction than a documentary.[24]

Another particularly harrowing episode occurred twenty-one years later, during the Reagan Administration and that of Yuri Andropov in the USSR. The details have been kept under wraps for decades, and even now much is classified, with the Russians being particularly close-mouthed about their side of the story, most of which is known in the West from accounts by only one man, Oleg Gordievsky, a Soviet spy turned double agent.

Here is the short version: in 1983, the Soviets mistook a NATO war-gaming exercise as a prelude to the real thing and came very close to preempting the expected attack with one of their own. After all, widespread thinking at the time was that the only way to survive a nuclear attack was to beat the other side to the punch, thereby reducing the number of incoming explosions as much as possible. The result was that, once again, the world nearly stumbled into nuclear war, because (among other things) neither side had much confidence in deterrence.

US–Soviet tensions had almost maxed out in 1983. Ronald Reagan took office as a strenuous anticommunist whose rhetoric was often downright bellicose. Early during the Reagan years, many in the Soviet hierarchy were convinced the United States was preparing for a first strike. Threats, especially nuclear ones, have a habit of doing double duty; intended to prevent an attack (as well as to rally domestic support by aggressive posturing), they often raise the specter of a possible attack, deterrence be damned. Reagan was not just posturing. Once in office, he authorized development and production of the Trident II submarine-based missile, B-1 and B-2 bombers, neutron bombs, doubling of the US military budget, increased funding for blast and fallout shelters, and so forth. Moreover, the United States had been initiating a series of probes to test Soviet response times and capacities, with naval vessels attempting to get as close as possible to Soviet nuclear facilities and aircraft, periodically approaching Soviet airspace at high speeds, only to veer off at the last minute. The result was an increasingly paranoid atmosphere in the Kremlin.

During the late 1970s, the USSR had begun deploying intermediate-range SS-20 missiles that could reach Allied forces in Western Europe. In response, NATO elected to field its own Europe-based ground-launched cruise missiles and Pershing II ballistic missiles. The United States avowedly intended these actions to induce the USSR to withdraw its SS-20s; but, from the Kremlin's perspective, they constituted a serious threat as part of a possible decapitating first strike. After all, the Pershing IIs could reach Moscow in less than ten minutes. Commenting on the advanced age of the Kremlin hierarchy, the Vice Chairman of the US National Intelligence Council observed that these missiles could strike in "roughly how long it takes some of the Kremlin's leaders to get out of their chairs, let alone to their shelters."[25]

In 1981, the Soviets initiated Operation RYaN (Russian acronym for "Nuclear Missile Attack"), the most extensive and far-reaching espionage undertaking in their history. Its motto, according to KGB General Oleg Kalugin, was "do not miss the moment when the West is about to launch war." Soviet foreign agents accordingly began monitoring anything considered relevant to a surprise attack, which comprised 292 identified indicators, ranging from pizza deliveries at the Pentagon to the movements of key military and political figures, as well as recording traffic in and out of nuclear facilities themselves. Misunderstandings were rife, such as one time when the Pentagon lights were on all night, which suggested to the Soviet spooks an episode of feverish war planning, but was actually routine janitorial work. Soviet intelligence also reported, incorrectly, that US bases had suddenly been placed on high alert, coincidentally during the precise time frame Soviet planners had estimated would elapse between a decision to launch an attack and when it would actually be carried out.

Making matters worse, a few months previously, in May 1982, someone at the Defense Department had leaked a document to *The New York Times* that contained plans for conducting a six-month-long nuclear war, with orchestrated sequences of reloading missile silos that even included retaining enough warheads to continue fighting yet another war, or possibly to deter future attacks. In any event, whatever their intention, these announcements ratchetted up the tension in the Kremlin. So, the stage was set for some very dangerous misunderstandings. It got worse.

An enormous naval exercise was conducted during April and May 1983, which was a particularly crucial year, with three Pacific-based carrier battle groups simulating total war against the USSR. By design, US forces began probing the sensitivity (and restraint) of Soviet air defenses. Especially provocative was a mock bombing exercise in which US Navy aircraft actually flew over a Soviet military base on Zeleny Island, which the Soviets countered by flying over the Aleutians, and leading Premier Andropov to order that any subsequent overflight of Soviet territory be shot down. At the time, the United States was about to deploy its Pershing II missiles in Europe. And then, sure enough, in September of that same fateful year, Soviet radar reported what they believed was a US spy plane in Soviet airspace over Siberia.

An Su-15 interceptor was ordered to shoot it down. The intruder was no spy plane, but Korean Airlines (KAL) flight 007, a commercial airliner bound for Seoul that had gone off course and whose communications with the scrambled Soviet jet were garbled. There were 269 civilian passengers, all of whom perished, including 62 Americans (one being a sitting congressman who was chairman of the bellicosely anticommunist John Birch Society) and 22 children younger than the age of 12.

The West was outraged. President Reagan denounced Soviet "barbarity," and then things got even more dangerous. On September 26, three weeks after shooting down KAL 007, a newly installed Soviet early-warning satellite system sent an alert that a US intercontinental ballistic missile (ICBM) had been launched and was heading toward the USSR, followed by a possible four more. The mid-ranking duty officer of the early-warning command system, Lieutenant Colonel Stanislav Petrov, decided—on his own and counter to explicit military protocol—not to report this incident, which he judged to be a false alarm. And so it was. (Part of his reasoning was that a real first strike would have involved hundreds of missiles, not just a handful.)

Had he informed his superior officers, General Secretary Andropov et al. would have had just minutes to decide whether to "retaliate" and, given the tense state of

US–Soviet relations at that time, the outcome could easily have been disastrous. Petrov was nonetheless officially reprimanded for his nonaction. Shortly thereafter, he left the Soviet military and, in May 2017, died in relative poverty. Interviewed for the film *The Man Who Saved the World*, Petrov said this about his role in the incident, "I was simply doing my job . . . that's all. My late wife for 10 years knew nothing about it. 'So what did you do?' she asked me. 'Nothing. I did nothing.'" Petrov's "nothing" could have saved the world.

One might think that after this harrowing event things couldn't have gotten worse, but they did. As incredibly bad luck would have it, just six weeks after the Petrov close call, and at precisely the time that the USSR's intelligence services were desperately looking for early indicators that the West might be about to initiate a nuclear attack, NATO actually began to simulate just such an attack! An extensive war game, called Able Archer 83, began on November 7, 1983, mimicking a full-fledged nuclear assault on the USSR. It was nothing less than a full dress rehearsal of how a first-strike nuclear war would in fact be conducted. The formal exercise ended four days later; but, during that interval, all hell almost broke loose. Able Archer 83 involved not only a formal handoff from conventional to nuclear authority, it also simulated loading of nuclear warheads aboard strategic bombers, along with heightened alert status on the part of missile silo crews and forward-deployed medium-range missiles.

According to a 2013 analysis by the National Security Archive,

> The Able Archer controversy has featured numerous descriptions of the exercise as so "routine" that it could not have alarmed the Soviet military and political leadership. Today's posting reveals multiple non-routine elements, including: a 170-flight, radio-silent air lift of 19,000 US soldiers to Europe, the shifting of commands from "Permanent War Headquarters to the Alternate War Headquarters," the practice of "new nuclear weapons release procedures," including consultations with cells in Washington and London, and the "sensitive, political issue" of numerous "slips of the tongue" in which B-52 sorties were referred to as nuclear "strikes." These variations, seen through "the fog of nuclear exercises," did in fact match official Soviet intelligence-defined indicators for "possible operations by the USA and its allies on British territory in preparation for RYaN"—the KGB code name for a feared Western nuclear missile attack.[26]

Declassified material made available in late 2018 by the National Security Archive further confirms that the Soviet leadership had been genuinely alarmed that a US/NATO first strike was about to commence, an anxious mind-set that was particularly dangerous at the time, and remains a potential risk today. Thus, a previously secret article appearing in the February 1984 Soviet General Staff Journal *Voennaya mysl* [Military Thought] contained an article by Defense Minister and Politburo member Dmitry Ustinov, reminiscing about Able Archer 83 that "it was difficult to catch the difference between working out training questions and actual preparation of large-scale aggression."

Able Archer had also been preceded by a sudden increase in coded communications between Washington and London—exactly what the KGB anticipated would occur before a coordinated US–UK nuclear attack—whereas the flurry of diplomatic to-and-fro had actually concerned the US invasion of Grenada at the time, a Caribbean Island that was officially under the sovereignty of the UK's Queen Elizabeth.

As for the Soviet response, the US President's Foreign Intelligence Advisory Board conducted its own investigation in 1990, the results of which were only made publicly available in 2015, after a more than decade-long battle with the National Security Archive.[27] The review revealed that the USSR had conducted thirty-six intelligence flights, "probably to determine whether US naval forces were deploying forward in support of Able Archer 83," while at the same time, all civilian air flights in Eastern Europe were canceled, probably to facilitate use of emergency aircraft as needed. Soviet military activities at the time suddenly escalated to an "unparalleled scale" that included Soviet fighters in Poland and East Germany placed on high-alert status for the only time during the Cold War, and Soviet helicopters transporting nuclear weapons from their storage sites to missile launch pads. At the same time, Soviet spy satellites were reporting that NATO bombers, conspicuously armed with nuclear weapons, had been leaving their hangars; it was subsequently revealed that these were dummies, adding realism to the ongoing Able Archer exercise.

These and other events made the Soviet leadership jumpy indeed. According to double agent Gordievsky, Marshall Nikolai Ogarkov, Chief of the Soviet General Staff, monitored these and other provocative developments in a bunker away from Moscow. The President's Foreign Intelligence Advisory Board later concluded, "Soviet military leaders [had been] seriously concerned that the United States would use Able Archer 83 as a cover of launching a real attack."

At this time, the West had its own largely unsung hero who may also have saved the world. Leonard Peroots was Assistant Chief of Staff for US Air Force Intelligence in Europe during Able Archer 83. Peroots noted how, in response to NATO's war game exercise, Soviet military forces—most assuredly not playing games—had increased their alert. Peroots reasoned that if he ordered a correspondingly heightened real-world strategic response by NATO, this would have risked confirming the Soviets' worst fears, leading to additional escalations that might have become unstoppable. And so, like Petrov, he did nothing—and as with two months earlier, the world kept spinning 'round. Millions of lives hung by a Damocles-like thread, which was kept intact by the mindful inactions of two midranking officials: one, American and the other, Russian.

Poor deterrence! It doesn't get much respect. After all, it floats somewhere between aggression and appeasement, succeeding only when nothing happens. More often than we'd like to think, however, it has come down to a small number of courageous individuals to extricate us from its clutches and ensure that nothing much continues to happen. Like Wile E. Coyote, we have been running over a chasm. And like that manic cartoon creature pursuing an unattainable meal, we haven't been especially wily in our pursuit of security.

In his memoir *An American Life*, Ronald Reagan wrote of finally allowing himself to look at the dangers of nuclear war after the events of 1983, implicitly raising a critique of deterrence, although without specifically mentioning Able Archer or its quiet heroes:

> Three years had taught me something surprising about the Russians: Many people at the top of the Soviet hierarchy were genuinely afraid of America and Americans. Perhaps this shouldn't have surprised me, but it did . . . . During my first years in Washington, I think many of us in the administration took it for granted that the Russians, like ourselves, considered it unthinkable that the United States would launch a first strike against them. But the more experience I had with Soviet leaders and other heads of state who knew them, the more I began to realize that many Soviet officials feared us not only as adversaries but as potential aggressors who might hurl nuclear weapons at them in a first strike . . . . Well, if that was the case, I was even more anxious to get a top Soviet leader in a room alone and try to convince him we had no designs on the Soviet Union and Russians had nothing to fear from us.

The Cuban Missile Crisis and Able Archer 83 exemplify what engineers call multisystems failure—situations in which many independent bad events happen simultaneously or in sequence. Although ostensibly improbable, they obviously can happen and have happened. Although there are many clichés about history repeating itself, the brutal reality, as noted earlier, is that history is full of events that had never occurred before.

Others, in the nuclear realm, have involved single incidents, the most serious being false alarms in which computer malfunctions or unusual atmospheric conditions—in one notorious case a flight of geese; in another, an explosion in a gas pipeline interpreted incorrectly as an ICBM launch—have caused a sudden, false perception that one side is under attack and that nuclear deterrence has failed. By definition, inadvertent nuclear war cannot be deterred. Worse yet, deterrence can make inadvertent nuclear war more likely.

Later we examine some additional circumstances in which deterrence nearly collapsed. Here is an especially well-documented case. It occurred four years after the end of the Cold War, on January 25, 1995, and is known as the Norwegian Rocket Incident, or the Black Brant Scare.[28] US and Norwegian atmospheric scientists had launched a four-stage rocket designed to study the aurora borealis. Its northbound trajectory carried it briefly along the route outlined for Minuteman III ICBM launches targeting Moscow. It also attained an altitude of more than 900 miles, of the sort that suggested it could constitute a high-altitude nuclear attack designed to incapacitate Russian early-warning radar. Russian nuclear crews went to high alert and the famously alcoholic President Boris Yeltsin was handed the Kremlin's equivalent of the US presidential "football," containing nuclear launch codes. Maybe he was just sober enough (or a nonbelligerent drunk); in any event, he decided not to respond.

It turned out that Russian authorities had been notified ahead of time about this launch, but the information had not been conveyed up the command chain. Russian observers eventually ascertained that the rocket wasn't entering their airspace and was not a threat. Twenty-four tense minutes after it had been launched, it fell to earth, as planned, near Spitsbergen.

In January 2018, the following alarm was issued by the Hawaii Emergency Management Agency: "Ballistic missile threat inbound to Hawaii. Seek immediate shelter. This is not a drill." Not surprisingly, Hawaiians panicked. Fortunately, they were not in a position to "retaliate," presumably against North Korea. Also fortunately, this erroneous report was quickly corrected by the US military, because no national warning systems indicated such an attack, and, moreover, despite the high tension at the time between Pyongyang and Washington, DC, no one seriously thought that Kim Jong-un would order an unprovoked attack on Hawaii. But what if a comparable error had been reported to the North Korean dictator? The North's warning system is largely constructed of outmoded Soviet hardware and, as far as can be known, there is essentially no redundant, validating system that would show that an error had occurred. In addition, even if a subordinate were to recognize some sort of malfunction—whether mechanical or human based—who would be willing to take the blame, given that Mr. Kim has shown little hesitancy in executing people who have displeased him?

By virtue of the deterrence they provide—i.e., the threat they pose—nuclear weapons are undoubtedly effective . . . in threatening those that possess them, largely because, although they promise security, they actually do the opposite. Moreover, one of the pernicious effects of deterrence for some proportion of the general public is the false sense of security that it produces. Part of human nature may well be vulnerable to the seductive allure of immense explosive power, generating a sense of security where a perception of peril should be.

James Baldwin once argued that the role of art was "to prove, and to help one bear, the fact that all safety is an illusion."[29] Better safe than sorry is an old saw that, like most old saws, retains some sharp teeth. No one would dispute that being safe beats being sorry, and that in a dangerous world, erring on the side of caution is yet another wise maxim. But history alone makes it increasingly clear that relying on nuclear deterrence involves erring on the wrong side.

In a scene that regrettably is often cut from budget-conscious productions, Hecate, the goddess of witchcraft, appears in Act 3, Scene 5 of Shakespeare's *Macbeth*. She antedates Baldwin—not to mention nuclear deterrence—by pointing out that "security is mortals' chiefest enemy," pointing to how Macbeth's belief in his invulnerability will ultimately result in his downfall. Seeking security and believing that we have achieved it, whereas since 1945 we have been going in the opposite direction, may well be our "chiefest enemy" too.

One of the more insidious effects of nuclear deterrence is that it ostensibly influences the behavior of other countries (regardless of whether it actually does so), all the while seeking to operate below the radar of its own citizens. The United States is not alone in refusing to confirm or deny the presence of its own nuclear weapons, pretty much anywhere. As a result, and because they cannot typically be seen, smelled, heard, or touched, they tend to lack psychological reality, thereby—as their devotees intend—anesthetizing citizenry into ignoring their existence or taking them for granted, while unconsciously deriving satisfaction from their dimly acknowledged potency. The danger lies in how difficult it is to mount opposition to something that is not readily apparent and is thus not obviously threatening. Out of sight, out of mind makes it especially difficult for us to confront our chiefest enemy.

As a kind of Aesop's Fable for our deterrence-addled times, here is a reminiscence by famed British psychologist Nicholas Humphrey, from his *Bronowski Lecture*,

> When I was a child we had an old pet tortoise we called Ajax. One autumn Ajax, looking for a winter home, crawled unnoticed into the pile of wood and bracken my father was making for Guy Fawkes day. As days passed and more and more pieces of tinder were added to the pile, Ajax must have felt more and more secure; every day he was getting greater protection from the frost and rain. On 5 November bonfire and tortoise were reduced to ashes. Are there some of us who still believe that the piling up of weapon upon weapon *adds* to our security—that the dangers are nothing compared to the assurance they provide?

There definitely are such people, and they dominate the governments and especially the military of nuclear-armed countries. Confidence in deterrence-based security based on a logically indefensible global threat posture exemplifies what Italian political theorist Antonio Gramsci called "optimism of the spirit" overcoming "pessimism of the intellect." An optimistic spirit is admirable, but in the case of nuclear deterrence, it should be superseded by a realistic pessimism of the intellect, deriving from a clear-eyed understanding of the threats posed by deterrence itself—unless we wish to emulate the person who falls (or jumps or is pushed) off the roof of the Empire State Building and optimistically reassures himself on route, "So far, so good."

SELF-UNDERMINING THREATS

In honor of Ajax, here are some of the dangers generated by the doctrine of nuclear deterrence. It is bad enough that its threat has neither deterred nor even provided useful coercive power, but its extraordinary risks are even more discrediting, producing a kind of failed-threat-squared: not only does deterrence fail to achieve the results attributed to it, but also several of its components directly threaten catastrophe in themselves, raising the prospect of self-generated disaster, and thereby imperiling threatener and threatened alike. In short, the threat-based system of deterrence unavoidably undermines itself. Here's how.

First is the problem of credibility. As we've seen, the believability of threats is an issue even in the world of animal conflicts, because animals must deal with the prospect that the threatener might—or might not—have the physical wherewithal along with the behavioral inclination to back up a threat. If so, then best take the threat seriously. If not, then it's safe to disregard the bluff and bluster—as long as you are sure that it really is bluff and bluster, and if the payoff (access to food, a mate, nest site, and so forth) of ignoring the possible danger is worth the risk. From the threatener's perspective, it might be worthwhile to express threats even if you can't support them, but only if the likelihood is that your bluff won't be called. Recall the difficult situation of Harris sparrows, having been experimentally made to seem more competent than they actually were, and whose bluff was then called by other, more genuinely intimidating animals. Credibility is no less important in criminal law and penology; whenever threats are issued, the question arises whether they are real or just for show.

If threats are connected with serious consequences, credibility becomes a real consideration. But perhaps nowhere does it loom as large as in the world of strategic nuclear threats, where it poses a huge obstacle to deterrence, such that the effort to overcome the credibility problem becomes a problem in itself, which in turn increases the likelihood that the whole system will someday collapse catastrophically. Here is a simple nonnuclear thought experiment that models the problem.

Imagine that your child refuses to eat her spinach, so you threaten to blow up the house. Your credibility will be low. What to do? Your threat would be more believable if you proceed to stash sticks of dynamite throughout the building and wire them to a plunger in the dining room, but even then, it probably wouldn't work, and most psychologists (not to mention firefighters and first responders) would likely advise against such a child-rearing technique. A police officer armed with a backpack nuclear weapon would similarly have a hard time deterring a bank robber: "Stop in the name of the law or I'll blow us all up!"

Analogously, worry arose during the Cold War that defending Europe with nuclear weapons would destroy it, and so the claim that the Red Army would be deterred by nuclear means was literally an empty threat, not credible.

On the other hand, let's say you are trying to deter your son from chasing the cat. You might threaten that if he does it again, he won't be allowed to play with the cat for perhaps a day, a week, or whatever. That would be credible, and if he transgressed, you could—and probably should—carry it out, not only for the sake of his training (and the tranquility of the cat), but also to reinforce your own parental cred. If credibility is important when it comes to being a parent, it is at the very heart of deterrence, which has given rise to an array of doctrines and weapons designed to give credence to something that it inherently lacks.

"One cannot fashion a credible deterrent out of an incredible action," wrote JFK's defense secretary, Robert McNamara. "Thus, security for the United States and its allies can only arise from the possession of a range of graduated deterrents, each of them fully credible in its own context." [30] Such thinking led, in turn, to the notion of "flexible response," according to which a country should possess a range of military options, including a range of responses short of all-out mutually assured destruction. But this answer turns out to be equally loony and here is why: to make nuclear weapons credible, they must be made usable, and the more usable, the more likely they are to be used. And yet, the repeatedly cited justification for nuclear weapons is that they exist so as *not* to be used.

Not only is the threat of nuclear retaliation likely to be useless, doing so could be lethally counterproductive because of its worldwide effects (fallout; massive firestorms; ozone depletion; nuclear winter; the devastation of economies, of societies, and of global ecology), while also raising the possibility of yet another attack from the aggressor's remaining nuclear forces, which would doubtless be on highest alert after conducting an initial strike. In addition, given the ethical issues raised by a willingness to commit mass murder on the largest scale in human history, there might be further reason to doubt a nuclear state's willingness to do so. As former UK Prime Minister Harold Macmillan pointed out, it would be merely posthumous revenge. Understood by an adversary, this adds further incredibility to deterrence. Worse yet, as we shall see, it also generates pressure for a particularly dangerous escalation known as "launch on warning."

Here is one way to enhance credibility even if the action being threatened is so illogical that no sane person would do it, even in retaliation: be crazy. Or act like you are. A US Strategic Command report in 1995 titled "Essentials of Post-Cold War Deterrence," argued the following:

> The fact that some elements [of the nuclear command authority] may appear to potentially "out of control" can be beneficial to creating and reinforcing fears and doubts within the minds of an adversary's decision makers. This essential sense of fear is the working force of deterrence. That the US may become irrational and vindictive if its vital interests are attacked should be part of the national persona we project to all adversaries.

Henry Kissinger was a little-known history professor when he was catapulted to prominence with his book *Nuclear Weapons and Foreign Policy* (1958), which sought to solve the dilemma of nuclear incredibility by, among other things, urging the development of weapons and tactics for fighting "limited nuclear wars." Easier said than done. A limited nuclear war is like being only somewhat pregnant or—more to the point—a little bit dead. Moreover, even without escalation to World War III, a limited nuclear war in Europe or between regional nuclear powers such as India and Pakistan, Israel and its adversaries, and so on, would appear altogether unlimited to the Europeans, South Asians, or Middle Easterners for whose benefit the war was ostensibly being fought. As a result, even limited nuclear war—hence, nuclear deterrence itself—lacks credibility after all. But nonetheless, deterrence remains official doctrine in every nuclear-armed country, even as they field weapons intended for "limited" battlefield use.

These and other problems to be elaborated shortly have led Lawrence Freedman, Professor of War Studies at King's College, London, to note that "the Emperor Deterrence has no clothes, but he is still Emperor." There is a quip favored by some strategic analysts that nuclear deterrence does not work in theory, only in practice. It is painfully clear that it really does not work in theory. Whether the naked emperor works in practice seems equally doubtful.

The emperor received what ought to have been its death blow during the early 1980s when researchers, including astrophysicist Carl Sagan, reported on the likely effect of nuclear war on global climate. Of special concern was the impact of firestorms of the sort produced by the atomic bombing of Hiroshima in 1945, which lofted enormous amounts of dust and—most important—soot into the upper atmosphere. Smoke particles can be expected to absorb solar light and heat, thereby generating an extended period of dark and cold down below. How dark and how cold depends on a number of factors, such as the size of the firestorms produced, the season in which the events happen, how high into the stratosphere the soot extends, and so forth. But the empirically based computer models were quite clear, and the phrase "nuclear winter" was born.[31]

The first term of the Reagan Administration combined a massive nuclear buildup with an extraordinarily cavalier attitude toward nuclear war and lip service devoted to surviving and even winning World War III. As a result, nuclear anxieties were ratcheted up to a level at least as high as worries are today about global climate change via greenhouse warming. Only in this case, the concern was catastrophic cooling, with estimates that summertime temperatures would plummet to the range normally occurring within each location's winter season. The duration of this post-holocaust deep freeze would depend on the rate at which the smoke particles eventually settled out of the atmosphere—a process liable to be very slow because such particles are typically microscopic and, moreover, during the process of blocking the sun's radiation, they would have been heated and caused to rise rather than fall.

Regardless of the specific parameters, one result, according to credible estimates, would be massive worldwide famine—on top of the devastating immediate impacts of blast, burns, and radiation that were more widely known at the time. Attacks on the nuclear winter research and on the reputation of the researchers began almost immediately, only in part a result of predictable skepticism by scientific competitors, but mostly because the nuclear winter scenario threatened to cut the legs out from under strategic war planning. After all, insofar as nuclear war would have catastrophic worldwide consequences, it could not be won in any meaningful sense, and might well not be survivable, even if the United States were, miraculously, to avoid immediate devastation. In fact, the research showed that the US and USSR would have no choice but to be *self-deterred*—an inconvenient truth for nuclear deterrence, which was supposed to act on the *other* side.

Initially, the Reagan Administration initially accepted the nuclear winter research, but responded by doubling down on how to avoid it. Assistant Secretary of Defense Richard Perle testified to Congress, "The Administration accepts that a nuclear exchange would produce a nuclear winter effect," before arguing, "Rather than eliminating nuclear weapons, the most realistic method of preventing nuclear winter is to build enough to make sure that the Soviets will be deterred from attacking."[32] Changing tune, there followed a series of publications (many of them sponsored by the Pentagon and right-wing think tanks) during the 1990s seeking to debunk nuclear winter itself, labeling it a hoax and simply "bad science." Carl Sagan was especially vilified, although for the most part both the Pentagon and the civilian nuclear priesthood denied the legitimacy of the research, when not ignoring it altogether.[33]

In fact, it was very good science, subsequently confirmed by a number of peer-reviewed studies showing that, if anything, earlier predictions were underestimates and that even the detonation of one hundred atomic bombs in a war between India and

Pakistan would result in roughly two billion deaths as a result of worldwide climatic changes alone, notably impacts on food production and distribution.[34]

Undeterred by these findings—or perhaps preferring "alternative facts"— deterrence has continued to march on, as though postnuclear climate science is as irrelevant to strategic policy as right-wing climate change deniers treat the impact of anthropogenic greenhouse gas emissions. "You've got to be honest," said comedian George Burns. "If you can fake that, you've got it made." Threats—especially nuclear ones—must seem real to be effective, but when it comes to deterrence, they have to be faked, precisely because they lack credibility, which is a big reason why we haven't got it made.

THE INCREDIBLE DILEMMA OF EXTENSION
AND ESCALATION

The doctrine of deterrence goes beyond its ostensible strategic role of preventing an attack on the homeland. In its incarnation of "extended deterrence," the so-called US nuclear umbrella is supposed to provide protective cover for certain crucial allies that do not have their own nuclear weapons but are said to shelter under the US arsenal. Attack these countries, goes the expectation, and the United States would retaliate as though its own soil had been struck.

In the first years after the Second World War, when—against vigorous protests from the USSR—the Western allies amalgamated their French, British, and US sectors into the unified state of West Germany, the Bonn government worried about the credibility of US extended deterrence. In 1956, the leaked Radford Plan (proposed by Admiral Arthur Radford, who was then Chairman of the US Joint Chiefs of Staff) called for the gradual withdrawal of most US battalions from Europe and an increased reliance on nuclear weapons—which were less expensive on balance, offering "more bang for the buck"—to defeat any Soviet invasion. But when the USSR shortly developed ICBMs that threatened the United States, US allies in Europe began to doubt that Washington would risk New York City to save Frankfurt or Hamburg. The French government in particular, under the very nationalistic Charles de Gaulle, felt that it needed its own *force de frappe nucleaire* to have a credible deterrent.

From the perspective of the United States and NATO, this emphasized the need for limited war-fighting options. NATO commanders had worried that the towns in Germany were only "two kilotons apart," which, in their view, necessitated smaller nuclear weapons as part of the capacity to make nuclear war-making a credible threat. Without such realism, there would be no threat that they would be used—and, without a credible threat, no deterrence.

The result was the elaboration of smaller, more accurate tactical weapons, including "neutron bombs," which were designed to kill people but leave structures relatively intact. They would, in theory, be more usable on the battlefield, and thus using them in a crisis would be more believable—but also necessarily more likely. So far, no nuclear strategist—no matter how brilliant or how immersed in political, mathematical, or military arcana—has been able to solve this problem. Unsolved, it is a problem indeed.

In fairness to those nuclear strategists who argue otherwise, a brief pause is warranted here. After all, there is another side to this argument, and in developing the antinuclear perspective, the current treatment has fallen into the trap of those sarcastically lampooned by the poet Ogden Nash, who achieved "the notable feat of one-way thinking on a two-way street."ˣ The other way on this particular street goes as follows: insofar as it lacks an adequate arsenal of small, accurate nuclear weapons designed for battlefield use, the United States could theoretically find itself faced with an opponent (presumably Russia) that believes it can intimidate its rival by escalating to a level such that a devastating strategic assault on the Russian homeland is the only option available to the US—one that, because it would be not only morally intolerable, but also it would also lead to an equally devastating counterresponse, could not be exercised.

Accordingly, it is necessary to have the capacity to match any battlefield escalation, including crossing the nuclear threshold, to prevent an opponent from doing so. In short, to avoid nuclear war, get ready to fight one.[35] This thinking ignores the fact that there is no battlefield effect to be achieved by tactical nuclear weapons that could not be accomplished via conventional weapons. And, of course, it also leaves unanswered the dilemma that by deploying weapons that are more usable, both sides are unavoidably betting their future on devices that are more liable to be used. You can't have it both ways. And considering the pros and cons of each, the cons of usability far outweigh the pros.

There is yet another problem, broached—although not solved—by theorists scrambling to find a way of detonating nuclear weapons during a conflict, but then stopping things short of all-out mutual destruction. Let's imagine that war has broken out and it has gone nuclear, at least on one or more battlefields, and "merely" at a tactical level. How to control the next step, and the ones after that? Herman Kahn, who never met a nuclear war scenario he didn't like, posited forty-four distinct rungs on an "escalation ladder," starting with "sub-crisis maneuvering," running through such precisely fine-tuned events as "non-lethal demonstrative detonations," "tactical strikes on military forces," and "small-scale attacks on civilians," ending with thermonuclear Armageddon, or as Kahn called it: "wargasm."[36]

ˣ From his poem "Oh Stop Being Thankful All Over the Place."

It never seriously occurred to this self-styled thinker of the unthinkable, celebrated for his hard-headed realism, to question whether these various steps could be choreographed like a Nijinsky ballet, how—aside from the problem of making clear-minded decisions—leaders would even keep accurately apprised of the damage their own side had experienced as well as imposed, how the other side's intentions would be comprehended, and how fully alerted, spring-loaded retaliatory missiles would be kept operational in the midst of "limited nuclear wars" for the next precisely calibrated escalation.

Kahn also introduced the concept of "escalation dominance," whereby the United States should always maintain the ability to exceed an opponent at every rung, thus projecting sufficient threat that said opponent would not seriously contemplate ascending to the next level. It has also been suggested—although not confirmed—that Soviet strategy (believed to have been inherited by today's Russia) has long involved readiness to "escalate in order to deescalate." The idea is that, by consistently moving up a step, Russia would signal its seriousness, thereby inducing the United States to step down, at the same time that US strategy is to engage in escalation dominance—in other words, each would somehow get *the other side* to step down by credibly threatening to step up!

As Deputy US Defense Secretary Robert Work pointed out in congressional testimony: "Anyone who thinks they can control escalation through the use of nuclear weapons is literally playing with fire. Escalation is escalation, and nuclear use would be the ultimate escalation."[37] Not that it's easy to fall off the ladder; it's nearly impossible *not* to. As one of our premier defense scholars has put it:

> If and when the nuclear use threshold is crossed, the most compelling need in the following few minutes would be to control escalation. If proponents of nuclear deterrence cannot address the issue of escalation control—and they have not even begun to explain their thinking on this topic, since to do so would invite derision—then there is no way that the battlefield use of nuclear weapons can be justified under the precepts governing a just war. There is no justice in the blast, fire, radiation and other nuclear weapon effects. There are only victims.[38]

In short, the scramble for credibility bumps yet again into its own incredibility. The option of small nuclear weaponry remains psychologically attractive, however, because its advocates maintain that such use would be less likely to evoke a full-scale strategic response from an adversary. Moreover, it would provide leaders with the option of responding to an attack without necessarily killing a hundred million or so of the other side's civilians right off the bat. Yet as we have seen, there are no plausible scenarios for how such use would *not* lead to escalation by the other side.

One worrisome suggestion of how to overcome the dilemma of making an incredible threat credible is the following prescription for effective brinkmanship from Thomas Schelling, a Nobel Prize-winning economist and one of the founding fathers of deterrence theory. In *The Strategy of Conflict*, Schelling describes brinkmanship as a means whereby one side coerces another, despite the fact that it cannot credibly threaten nuclear war. Brinkmanship, for Schelling, is "the deliberate creation of a recognizable risk of war, a risk that one does not completely control. It is the tactic of deliberately letting the situation get somewhat out of hand . . . . It means harassing and intimidating an adversary by exposing him to a shared risk [of nuclear war]."[39]

In an earlier book, *Arms and Influence*, Schelling had first proposed his model, calling it "the threat that leaves something to chance."[40] He suggested that we imagine two climbers roped together near the edge of a crevasse, where they're arguing about something (presumably something *very* important). One of them, it doesn't matter which or why, wants the other to bend to his will. If he simply announces his demand, the other could simply ignore it. Imagine further that one wants to coerce the other by threatening to jump, in which case both would die; remember, they are roped together. Such a threat would lack credibility. But suppose the threatener moves right to the edge, becoming not only more insistent but also increasingly erratic in his footsteps. Perhaps he starts leaping up and down and shuffling his feet wildly. As a result, his credibility would be enhanced, not because jumping would then be any less suicidal, but because by increasing the prospect of shared calamity, emphasizing his threat by adding a soupçon of potentially lethal unpredictability, the threatener's literal brinkmanship just might kill them both. After all, chance factors—a sudden loss of balance, a gust of wind—might do what prudence would otherwise resist. Thus would unpredictability, according to Schelling, surmount the problem of incredibility. This terrifying loss of control wouldn't be a bug, but a feature. But the downside of leaving something to chance (something, moreover, that is pretty important) is that it leaves that thing to chance!

Schelling's thought experiment may have been inspired by the word "brinkmanship," which seems to have been first used by Democratic Party presidential candidate Adlai Stevenson, who, at a campaign event in 1956, criticized Republican Secretary of State John Foster Dulles for "boasting of his brinkmanship—the art of bringing us to the edge of the nuclear abyss."[41] At about this time, Henry Kissinger began developing the concept of the "security dilemma," in which "The desire of one power for absolute security means absolute insecurity for all the others."[42] Describing an early case of the security dilemma, Greek historian Thucydides wrote of the Peloponnesian War, "What made war inevitable was the growth of Athenian power and the fear which this caused in Sparta."

The idea of security dilemmas has typically been applied to the problem of reciprocal arms races, whereby a country's effort to counter a perceived military threat by building up its arsenal results in its rival feeling threatened, which leads that rival, in turn, to build up its arsenal—and so on. As a result, both sides end up less secure than they were before. Brinkmanship, à la Dulles and Schelling, introduces yet another dilemma: when a lack of credibility leads to various stratagems intended to enhance it, and that may well succeed in doing so, but in the process reduces security on all sides.

IRRATIONAL RELIANCE ON RATIONALITY

If nuclear deterrence were undermined by the problem of credibility alone, that would be bad enough; but, there are other skeletons in its closet. For one, deterrence relies on the assumption that all parties to this particular ghastly dance are, as pretty much all nuclear strategists assume, deeply rational, with both threatener and threatened locked in a reciprocating relationship in which each assumes both roles simultaneously. Hence the analogy suggested by Dr. Oppenheimer, that they are like "two scorpions in a bottle," but with the assumption that each will calculate gains and losses, coming up with similar results so that neither one stings the other. But, as psychologists know, this assumption itself is flawed—and not just for scorpions. If we know anything about human nature, it is that we are complex and unpredictable creatures, not logic-chopping computers.

Just as it makes little sense to assume that murderers are so rational as to be deterred when threatened by the prospect of capital punishment, there is little basis to assume that national leaders—especially when in the clutches of an anxiety-ridden situation—will reach into their rational selves and be deterred from acting in a way that is intellectually untidy and inconsistent with logic.

"Only part of us is sane," wrote Rebecca West. "Only part of us loves pleasure and the longer day of happiness, wants to live to our nineties and die in peace . . . ."[43] It requires no arcane wisdom to realize that people often act out of misperceptions, anger, despair, insanity, stubbornness, revenge, pride, and/or dogmatic conviction—particularly when under threat. Moreover, in certain situations—as when either side is convinced that war is inevitable or under pressure to avoid losing face—an irrational act, including a lethal one, may appear appropriate and even unavoidable. When he ordered the attack on Pearl Harbor, the Japanese Defense Minister observed, "Sometimes it is necessary to close one's eyes and jump off the Kiyomizu Temple" (a renowned suicide spot). During the First World War, Kaiser Wilhelm wrote in the margin of a government document, "Even if we are destroyed, England at least will lose India." While in his bunker during the final days of the Second World War, Adolf Hitler ordered what he hoped would be the total destruction of Germany, because he felt its people had "failed" him.[44]

Boris Yeltsin, president of the Russian Federation from 1991 to 1999, was a known alcoholic who became incoherent and disoriented when on a binge. Nothing is known about what contingency plans, if any, were established within the Kremlin to handle nuclear crises when they arose during the Yeltsin period. Richard Nixon also drank heavily, especially during his stressful stint as US President during the Watergate crisis, which ultimately led to his resignation. At one point, Nixon announced, ""I can go into my office and pick up the telephone and in twenty-five minutes seventy million people will be dead.""[45]

During that time, Defense Secretary James Schlesinger took the extraordinary step of insisting that he be notified of any orders from the President that concerned nuclear weapons *before* they were passed down the command chain. Presumably, Schlesinger— and by some accounts, Kissinger, who was then serving as National Security Adviser— would have inserted themselves, unconstitutionally, to veto nuclear war if Nixon had ordered it. As civilians, neither Yeltsin nor Nixon—when incapacitated by alcohol— would have been permitted to drive a car, yet they had full governmental authority to start a nuclear war by themselves.

In 1973, at the time that Nixon was under intense emotional pressure because of the Watergate investigations, Harold Hering, an Air Force major who had served several tours as a helicopter pilot in Vietnam and been awarded a Distinguished Flying Cross for his heroism, was in training to become a Minuteman ICBM launch control officer.

"How can I know that an order I receive to launch my missiles came from a sane President?" he asked his instructors.

Major Hering also inquired, more generally, "What checks and balances exist to verify that an unlawful order does not get in to the missile men?"[46]

In other words, what assurance would he have that a launch order would be legal under military law. He never received an answer; he was immediately removed from his training program and then, following a military review board hearing, was summarily discharged from the Air Force. Consider, à la Major Hering, a US president who shows clear signs of mental illness—notably, malignant narcissism combined with a dose of aggressive paranoia—inability to tell truth from falsehood; violent irritability plus intellectual laziness, impulsive anger, and verbal aggressiveness, vindictiveness, and sociopathy; added to a lack of curiosity, chaotic decision making; and someone whose statements and tweets have also been frighteningly consistent with dementia or genuine psychosis (notably, bipolar disease). Following the 2016 election, as it has always done since 1945, the United States handed over control of its nuclear weapons to this man, who, according to a warning letter signed by fifty leading *Republican* experts in national security, exhibits dangerous qualities in an individual . . . with command of the US nuclear arsenal. . . . is unable or unwilling to separate truth from falsehood . . . lacks self-control and acts impetuously . . . does not encourage conflicting views [and] cannot tolerate personal criticism. He has alarmed even our closest allies with his erratic behavior.

National leaders, nuclear armed or not, aren't immune to mental illness. Yet, keeping the nuclear peace via the carefully balanced use of threats presumes otherwise, wagering that those with their fingers on the nuclear trigger are rational actors who will also remain calm and cognitively unimpaired under extremely stressful conditions—a substantial challenge, even for those who are emotionally and intellectually competent. It also presumes that all leaders, regardless of their mental fitness, will always retain control over their forces and that, moreover, they will always retain control over their own mental faculties as well, making decisions under intense threat based solely on a cool calculation of strategic costs and benefits. The core idea of deterrence is that each side will scare the pants off the other with the prospect of the most hideous consequences, whose leaders will then conduct themselves with the utmost in cool, deliberate, rationality. Virtually everything known about human psychology suggests this is absurd.

In his book, *An American Life*, written after his two terms as president, Ronald Reagan recounted how his initial enthusiasm for nuclear deterrence had been extinguished when he realized that the enterprise could require making planetary life-or-death decisions in a time frame so short as to preclude rational thought: "The Russians sometimes kept submarines off our East Coast with nuclear missiles that could turn the White House into a pile of radioactive rubble within six or eight minutes. *Six minutes* [emphasis in original] to decide whether to unleash Armageddon! How could anyone apply reason at a time like that?"

Donald Trump as President of the United States has been a one-man argument for nuclear abolition, and although the threat posed by nuclear weapons and the posture of deterrence may be ameliorated by putting them into more rational and thus more reliable hands, it would not be solved. When the nuclear codes were transferred as he was sworn in, the unsuitability of this situation became painfully evident, although as Ban Ki-moon often noted when Secretary General of the United Nations, "There are no right hands for wrong weapons." Dr. Ira Helfand, a leading figure in the physicians' antinuclear movement,[xi] has argued,

---

[xi] This includes the International Physicians for the Prevention of Nuclear War, which was awarded the 1985 Nobel Peace Prize, as well as the US counterpart, Physicians for Social Responsibility.

[T]hat the US has transferred its entire nuclear arsenal into these unequivocally wrong hands is an opportunity for us to refute the whole idea of deterrence as an acceptable nuclear policy. That policy assumes that the great power arsenals will be controlled by rational, responsible people who can be deterred. Many who have not previously agreed with us that there are no right hands, do understand how insane it is to have Trump in charge of the arsenal, and it can and should be a teachable moment when we get them to understand that the weapons are too dangerous and need to be abolished. You can't leave a loaded gun around where children can find it, and if you can't guarantee there won't be children around, you have to make sure there is no gun for them to find. The child is here; the gun has to go.[47]

Despite the widespread progressive revulsion toward Trump, an additional narrative has also emerged: that he isn't the problem so much as a symptom of a greater one, manifested in the rise of populist, nationalist authoritarianism worldwide. This is true enough, and it is also true that, even without such leaders in office, nuclear weapons and the doctrine of deterrence that justifies them would nonetheless be unacceptably dangerous. But in many respects, notably when it comes to the immediate danger posed by unstable leaders with control of nuclear weapons, they also *are* the problem.[xii]

As Helfand pointed out, a gun is dangerous in anyone's hands, but worse yet when wielded by a child (perhaps especially a "grown-up child"), a psychotic—even one who psychiatrists might describe as "well compensated"—someone liable to act out of whim, anger, or ignorance, unmediated by the restraining influence of others, and/or lacking in empathy for prospective victims or a clear understanding of the consequences of their actions. No one is so wise, so judicious, so reliable and trustworthy that the world should trust him or her with the capacity to end civilized life, and quite possibly much of life altogether. In short, *no one* should have the authority to launch a nuclear war. Not now; not ever—which makes continued adherence to deterrence especially ill-advised.

The presumption of consistent and enlightened self-interest is also the foundation of the now-discredited "rational market" theory of economics. For many years, economists—especially from the Chicago School (Milton Friedman et al.)—claimed

---

[xii] Following a briefing intended to showcase for the newly elected president successful arms control agreements that had led to a reduction in the US nuclear arsenal from its 1960s high, Trump reacted by asking, petulantly, why he didn't have as many nukes as his predecessors had. When he repeated this complaint during a later meeting, this led then-Secretary of State Rex Tillerson to observe that the president was "a fucking moron." (Recounted in Fred Kaplan's *The Bomb: Presidents, Generals, and the Secret History of Nuclear War*, Simon & Schuster, 2020).

that, overall, US stock prices reflected real wisdom because, in the aggregate, investors make rational decisions. And, moreover, that people maximize their "utility" by acting to enhance their personal well-being on a daily, even minute-by-minute basis. But psychologist Daniel Kahneman received a Nobel Prize in 2002 for his work demonstrating how the human mind creates "heuristic biases" that reinforce some beliefs and reduce others, often counter to rationality. In his book, *Thinking Fast and Slow*, Kahneman demonstrated that rationality is a very tenuous thing, in which "system 2," which involves calm, deliberate, rational thought, is typically preceded (and often superseded) by the instinctive and rapid-fire "system 1."

Many people do not need intense fear or obvious psychosis to behave in ways that are irrational, vengeful, even spiteful and self-destructive (i.e., hurtful to themselves as well as others). Look at school shooters and other terrorists who plan their own suicide along with random deaths. The World Health Organization noted in 2017 that depression was the number-one cause of disability worldwide, and it certainly is not realistic to insist that national leaders are immune to mental illness or drug abuse. Moreover, they may be the victims of insufficient or faulty information, or of other perceptual distortions that cause them to make incorrect judgments of others' intentions, the likely outcomes of various alternative courses of action, and so on.

Betting the future on the cognitive and emotional functioning of a few men—and in most cases they are, in fact, men—under high stress seems, at best, a dubious wager. In his poem, "Sunday Morning," Wallace Stevens wrote that "we live in an old chaos of the sun/ or old dependency of day and night . . . ." We also live in a new chaos of nuclear deterrence, dependent on who-knows-what (day, night, luck, God, whim, or whatever) for survival.

COUNTERFORCE'S COUNTERLOGIC

Beyond the problems of credibility and reliance on strict rationality, other frightening incongruities lurk behind the mask of deterrence. In a gun-toting society, a rational response to gun violence is to take away the guns. Until such time as the world gets rid of nuclear weapons, it might appear sensible, accordingly, for a country to seek the ability to disarm a would-be opponent. The technical term is "counterforce," and in the twisted world of nuclear deterrence, it is anything but sensible. Here, then, is yet another skeleton in the closet of deterrence.

Strategic deterrence by the threat of retaliation requires that each side's arsenal remains invulnerable to a first-strike attack, or at least that such an attack would be prevented insofar as a potential victim retained a "second-strike" retaliatory capability sufficient to prevent the attack in the first place. Over time, however, nuclear missiles have become immensely more accurate, raising concerns about the vulnerability of these weapons to a "counterforce" strike—one that targets a potential adversary's nuclear forces for destruction. In the perverse argot of deterrence theory, this is called "counterforce vulnerability," with "vulnerability" referring to the target's nuclear weapons, not its population.

A notable book, now several decades old, that described the early theorists behind nuclear deterrence was titled *The Wizards of Armageddon*. Toward the end of a similarly titled book, *The Wizard of Oz*, the alleged wizard confesses, "I'm really a very good man; but I'm a very bad Wizard." Many deterrence theorists are relatively good men (once again, nearly all are male), but they really are very bad wizards. A key rule of their fellowship requires that each side's arsenal remain safe from attack, or at least that the probability of them being destroyed in a first strike must be very low. Over time, however, as nuclear missiles have become extraordinarily accurate, concerns have risen about the growing vulnerability of these weapons.

First, a quick bit of history. When the United States had a nuclear monopoly, its official deterrence strategy was "massive retaliation"—an explicit threat that the US Air Force would wipe out the USSR in the event of any Soviet aggression. Then, after the USSR developed not only atomic bombs and hydrogen bombs, but also ICBMs, the strategy moved to mutually assured destruction. Through the late 1950s, essentially all of the 1960s, and even into the 1970s, there was only one basic option in the event of war with the USSR: all-out devastating nuclear attack, which would devastate not only the Soviet Union and the countries of Eastern Europe (East Germany, Poland, Czechoslovakia, Hungary, Romania, and Bulgaria), even though their population had no responsibility for what the Kremlin might have done, and also all of China, even if that country was otherwise unconnected to the hostilities.

In the United States, interservice rivalry played a major role. Early on, the Navy brass resented the Air Force's nuclear weapons monopoly, to the point that a parade of admirals testified to Congress that the prospect of atomic bombing, with its vast civilian casualties, was immoral and counter to American values. Meanwhile, the Air Force staunchly argued for the wisdom of mutually assured destruction—that is, targeting Soviet cities, which was pretty much all that comparatively inaccurate strategic bombing could do.

Then came submarine-based Polaris missiles, whereupon the Navy changed its tune, becoming fervent nuclear advocates. Air Force commanders—who had long possessed an unquenchable yearning to prepare for all-out "general war" against the USSR, complete with obliteration of its civilian population—accordingly began advocating for counterforce as a counter to the Navy, expressing dismay at the continuing inaccuracy of Polaris, as well as its successor, Poseidon missiles. The Navy's imprecision, the Air Force pointed out, was suitable for deterrence via mutually assured destruction (which the Air Force had earlier advocated when their bombers and early-generation

ICBMs were unable to accomplish anything else), but could do nothing but demolish cities because they were less accurate than the Air Force's newly deployed Minuteman II land-based ICBMs. So, the Air Force lobbied hard for counterforce, pretty much reversing its earlier commitment to "countervalue" targeting.

But technology marched on. With its several generations of highly accurate Trident missiles, the Navy soon caught up with the Air Force, whereupon the two competing service branches both agreed at last on the need for counterforce. Until then, the United States had largely been running the nuclear arms race with itself. The Soviets, however, were not altogether irrelevant and they, too, although consistently behind the US—especially when it came to missile accuracy—eventually caught up. (They tended to compensate for lesser accuracy by deploying more and larger warheads on larger missiles.)

The evolution of Soviet/Russian nuclear strategy is still not very well known. We do know that in the United States, counterforce was publicly promoted as a way of limiting a nuclear attack to the USSR's military targets rather than its cities, thereby, it was claimed, making it more likely that any Soviet retaliation would similarly refrain from devastating the US population in return.

Today's counterforce doctrine thus represents a third incarnation of deterrence theory, moving from massive retaliation to "countervalue" (targeting cities as such[xiii]) to "counterforce," targeting an opponent's weapons—missiles, submarine and bomber bases, arms depots, command and control centers, and so forth—supposedly to provide a greater variety of nuclear options while also attempting to limit unnecessary carnage on both sides. It was formalized in 1979 during the Carter Administration, under what is known as Presidential Directive 59. Tellingly, it was a successful initiative from the Pentagon alone, coming as an unwelcome surprise to the US State Department.

If a policy of capital punishment in criminal cases called for the execution of a lawbreaker's family, most people would protest against its immorality. Superficially, counterforce seems to have the same aura of ethical legitimacy as does killing only the killer. Better yet, it conveys the image of a "missilized" Lone Ranger, heroically and morally shooting the pistol out of a bad guy's hand, with no "collateral damage"—not even to the bad guy! However, aside from the impossibility of eliminating the other side's missiles without even spilling his cappuccino, counterforce targeting is profoundly dangerous, going counter to the avowed goal of deterrence because it raises concerns that the Lone Ranger may be planning a first strike. This is because, in theory at least,

[xiii] The "as such" is important, because essentially every large city is associated with numerous military targets, so whether the former is slated for destruction as "collateral damage" or whether it is directly targeted (as a population center) matters in the fine print of ethics and Just War theory, but not at all in practice. For example, Moscow was long targeted with more than sixty nuclear weapons because it housed, the Kremlin, KGB headquarters, and so forth.

a successful counterforce attack would preclude retaliation, thereby undermining deterrence. This works in several ways. For one, the capacity—even if just theoretically possible—to make a successful first strike (i.e., one that makes it impossible for the victim to retaliate) could very well motivate the more hawkish advisers and leaders to initiate such an action, especially in circumstances of high tension.

Counterforce has nonetheless been justified as serving deterrence, rather than providing a smokescreen for the development and deployment of first strike-weapons, which it may or may not actually be. In any event, although neither the United States nor Russia has ever announced publicly that its counterforce capability is being developed with an eye toward achieving a disarming first strike, both sides worry that this is in fact the *other's* unacknowledged intent. There is a fine line between deterrence and provocation, and for many (especially those potentially on the "receiving end"), counterforce crosses that line.

Making matters worse, in addition to (possibly) encouraging one side to initiate an attack if it believes that it possesses the counterforce capacity to preclude retaliation, counterforce weaponry makes such an initial attack more believable—more threatening—from the perspective of a would-be victim. And if a potential victim thinks that striking first could be appealing to the more rabid faction among prospective attackers, such knowledge risks generating a race to be the first to do so: if you are at risk of being disarmed, better to disarm the opponent first. In short, use them or lose them. The possibility that they *might* be lost raises alarm that deterrence could realistically fail and an attack be imminent, producing a dynamic that is especially troublesome during a crisis, when there is no limit to this chain of reasoning: side A, fearing that side B is about to preempt in this way, may be tempted to pre-preempt, leading side B, which anticipates such a pre-preemption, to consider pre-pre-preempting, and so forth—a potential death spiral for strategic stability.

This conundrum was recognized by thoughtful nuclear theorists as early as the mid 1950s, when the Soviet Union achieved its own comparatively inaccurate strategic nuclear arsenal (almost entirely bomber based at the time). Accordingly, the "wizards of Armageddon" emphasized the stabilizing benefit of maintaining a second-strike capacity: the ability to withstand a first strike and still retaliate. It appears that current theorists—at least, the ones with any clout in the Pentagon—don't realize or acknowledge that by perfecting counterforce, they are almost literally shooting themselves, and not just in the foot. Or, as famed negotiating guru Roger Fisher put it (personal communication), nuclear-armed countries are all in the same boat, with counterforce targeting reflecting each side's determination that it will be made more secure by making the other side more tippy. It is a counterlogical aspect of deterrence, however, that security comes from making one's potential opponents more secure, not less; counterforce works the other way.

In his book *Catastrophe Theory*, Russian mathematician Vladimir Arnold described the "fragility of good things," which he attributed to the fact that

> a small change of parameters is more likely to send the system into an unstable region than into a stable region. This is a manifestation of a general principle stating that all good things [e.g., stability] are more fragile than bad things. It seems that in good situations a number of requirements must hold simultaneously, while to call a situation bad even one failure suffices.

Bad things, too, can be unstable, and their instability is a major reason nuclear deterrence is a bad thing. At the same time, Arnold's principle holds for life itself, which most people consider both good and fragile.

Thomas Schelling, whom we encountered earlier via his Gedanken experiment regarding the crazily cavorting mountaineer, also offered this metaphor, italicizing how a posture of mutual threat increases instability:

> If I go downstairs to investigate a noise at night, with a gun in my hand, and find myself face to face with a burglar who has a gun in *his* hand, there is danger of an outcome that neither of us desires. Even if he prefers just to leave quietly, and I wish him to, there is danger that he may *think* I want to shoot, and shoot first. Worse, there is danger that he may think that I think *he* wants to shoot. And so on.[48]

Bringing it back to nuclear deterrence, Schelling warned about a country's military and political leaders engaging in ever-narrowing cycles of getting into each other's head: "He thinks . . . ; we think he'll attack. So he thinks we shall. So he will. So we must." Bear in mind that this warning came from a brilliant thinker and strategist who was one of the founders of nuclear deterrence theory, and hardly a peacenik. It would seem obvious that targeting the other side's supposed deterrent is a very bad idea, and yet the momentum for doing so is almost unstoppable. It isn't clear what—beyond interservice rivalry and ostensible ethics—drives this march to undo the supposed logic of deterrence. There are many additional factors; in the language of psychoanalysis, counterforce appears "overdetermined."

In addition, there is the intuitively persuasive idea that the more of the other side's nuclear weapons you destroy, the fewer are left to destroy you—notwithstanding that beginning an orgy of destruction in the expectation of limiting it would almost certainly begin it. Closely related is the understandable desire by political and military planners alike to have a range of possible options "if deterrence fails," even though having such a range generates the paradox of increased likelihood that they will be used, thereby causing deterrence to fail.

Don't underestimate, as well, the persistent military demand for accuracy, with its centuries-old insistence on target practice. Threatening to obliterate a city doesn't require accuracy. One or a small number of enormous air bursts would easily do the trick, no matter how poorly aimed. But destroying an ICBM that sits beneath immensely thick reinforced steel and concrete lids requires, if not pinpoint accuracy, at least a degree of precision that is consistent with hallowed military tradition.

Moreover, it qualifies as "technically sweet," a phrase often used by engineers and physicists to indicate something that is satisfyingly complex yet also achievable. In addition, this technical sweet-tooth feeds a gnawing financial appetite on the part of what Dwight Eisenhower labeled the military–industrial complex. After all, a counter-force capability not only requires scientific sophistication, but also engages the world of high-tech design and production, suggesting that its fundamental drivers may be no different from those economic and political factors that have long operated when it comes to conventional military procurements.

When it comes to financial and career incentives, counterforce targeting offers yet another payoff. Russia has 12 cities with more than 1,000,000 people, and barely 200 with between 100,000 and 1,000,000, each of which could be obliterated with one or two nuclear explosions. Even assuming two per city (just to be "safe"), pure deterrence in the form of countervalue targeting simply doesn't require more than a few hundred missiles, bombers, warheads, and bombs. But if the goal is to disarm—i.e., attack successfully—all of the other side's nuclear weapons, along with storage and support sites and command-and-control facilities, then potential targets are in the many thousands. With a counterforce strategy, opportunities almost literally explode for civilian jobs and money, along with career advancement within those military branches tasked to accomplish the mission.

This book does not attempt to examine the complex and extensive connections between financial gain and the organized enterprise of counterforce-based deterrence. Here is just one tidbit. Every year the Nuclear Deterrence Summit is held in Arlington, Virginia, just outside Washington, DC. According to its website, the 2019 meeting was

an opportunity for heavy hitters in government and the military to Network with their counterparts in industry . . . [and to] gather information through conference sessions, such as deterrence outlook, modernization strategy, updates from the National Nuclear Security Agency labs, Congressional initiatives and more! . . . as well as to Network with colleagues and government officials, 40% of whom are senior management executives (VP-level and higher) and 43% are government or military executives.

The organizers go on to extoll opportunities to "create business" for the next year "and beyond." Issues addressed by the various speakers included "modernization of the nuclear deterrent," an especially interesting phrase because it reifies the justifying label "the nuclear deterrent." That is, whereas deterrence refers to a process, "the deterrent" is taken to be synonymous with the weapons themselves, which are not merely expected to serve, bolster, or ensure deterrence—rather, they are *the deterrent*.[xiv] Deterrence itself is taken for granted, without debate. Case closed.

The 2019 conference was sponsored by Longenecker & Associates, AECOM, TechSource, Jacobs, Huntington Ingalls Industries, SOC, General Dynamics, Leidos, Fluor, Northrup Grumman, BAE Systems, Centerra, Honeywell, Bechtel, and others, most of them multibillion-dollar companies and all heavily invested in deterrence continuing. The registration fee for that event was $1,895 per person and $2,995 for exhibitors, which—considering the potential payoff to the participants—is likely a profitable investment.

It is bad enough that something so consequential is driven by such base considerations. At the same time, in the opinion of some of its more radical critics, things are more dire. It is at least possible that counterforce has evolved less in the context of somehow preventing an attack, than as a crucial adjunct to initiating one. If this seems paranoid, consider once again that to be effective, counterforce is indistinguishable from a first strike, given that both focus on destroying an opponent's nuclear weapons, despite the instability that such a capacity produces. After all, a first strike against cities that left the victim's nuclear weapons intact would pretty much guarantee catastrophic retaliation. Currently deployed, highly accurate missiles include the Trident D5, and a new generation of maneuverable gravity bombs is also being developed, despite substantial opposition by arms controllers.

More important than how they are seen by domestic critics, however, is how they are perceived by a potential victim, and the impact they have on the other side's planning and deployment, not to mention the unavoidable fact that during a crisis, their existence ratchets up tension and possible pressure to preempt, and in turn, to preempt any preemption ad infinitum. Being "firstest with the mostest" was an adage attributed (wrongly, it turns out) to Confederate General Nathan Bedford Forrest. But when two nuclear-armed rivals are each worrying that the other might go first, this maxim could be all too relevant.

Paradoxically, and counter to counterforce itself, stability under deterrence (insofar as it is possible at all) requires that a potential opponent be reassured, not threatened.

[xiv] This could serve as a classic example of fallacious reification, treating an abstraction as if it were real.

But reassuring a would-be adversary goes against the DNA of most military and political leaders, not only the most hawkish.

Alas, there is additional pressure for counterforce, beyond its possible first-strike appeal to some pronuclear planners—namely, the perception that it is more ethical than countervalue targeting. Which it is, if only because aiming at undefended civilians has long been judged immoral, even as it has, equally long, been done. When President Truman announced the atomic bombing of Hiroshima, he softened the horror by claiming—falsely—that the city was an important military base, whereas it was not. By August 6, 1945, Japan had no effective air defenses; hence, every conceivable strategic site had been bombed—many times over. It was subsequently revealed that Hiroshima, and shortly thereafter, Nagasaki, were selected precisely because, as civilian cities lacking strategic importance they had not been devastated by conventional explosives and firebombs, thereby providing a convenient laboratory to demonstrate the effect of a single atomic bomb.

The purported ethics of counterforce targeting represents the kind of faux morality that is yet another mask behind which the threat of deterrence resides. The redoubtable Thomas Schelling fomented something of a revolution in strategic thought, which had generally emphasized the importance of keeping suffering to a minimum, when he connected nuclear deterrence to the threat of causing pain to an opponent. Schelling argued that when it comes to nuclear weapons in particular, the ability to induce suffering (notably to civilians), is "among the most impressive attributes of military force." The announced goal, however, was not to inflict as much pain as possible, but to benefit from the *threat* of doing so. "The power to hurt is bargaining power," he wrote. "To exploit it is diplomacy—vicious diplomacy, but diplomacy."[49]

AN OXYMORONIC MORALITY

From the early stages of the Second World War, attempting to win by decimating civilians was a deliberate strategy. City bombing via massive aerial blitzes may have felt bloodless from the air, but it certainly caused pain on the ground. Tellingly, General Curtis LeMay, perhaps the most enthusiastic proponent of all-out city bombing, justified the firebombing of Tokyo on March 9 and 10, 1945 as follows: "There are no innocent civilians. It is their government and you are fighting a people, you are not trying to fight an armed force anymore. So, it doesn't bother me so much to be killing the so-called innocent bystanders."[50] LeMay had earlier planned the firebombing of Dresden as well. In these cases, pain is produced and not just threatened. Schelling's perspective is therefore benevolent by contrast: he argued, not for deterrence by punishment, but for deterrence through the *threat* of punishment.

The ethical and moral dimensions of nuclear deterrence have been addressed periodically, notably by theologians, but then typically relegated to afterthoughts. They are, however, stark. Philosophers are essentially unanimous that a nuclear war could never meet so-called Just War criteria. These include respecting noncombatant immunity, qualifying as a situation in which the overall peace-promoting benefits outweigh the costs, minimizing the amount of destruction inflicted, reflecting military necessity, and so forth. The Second Vatican Council therefore concluded that "any act of war aimed indiscriminately at the destruction of entire cities or of extensive areas along with their populations is a crime against God and man itself. It merits unequivocal and unhesitating condemnation." The Seattle Archdiocese of Archbishop Raymond Hunthausen has long included Submarine Base Bangor, home to a fleet of Trident submarines with their nuclear missiles.[xv] On June 12, 1981, Hunthausen delivered a speech in Tacoma, Washington, that was heard 'round the world and especially in the Vatican when he said, "We must take special responsibility for what is in our own backyard. And when crimes are being prepared in our name, we must speak plainly. I say, with deep consciousness of these words, that Trident is the Auschwitz of Puget Sound."

Adding to the longstanding doctrine that if something is immoral to do, then it is also immoral to threaten, the American Catholic bishops concluded in a pastoral letter that "this condemnation, in our judgment, applies even to the retaliatory use of weapons striking enemy cities after our own have already been struck." And in a message to the 2014 Vienna Conference on "The Humanitarian Impact of Nuclear Weapons," Pope Francis further declared, "Nuclear deterrence and the threat of mutually assured destruction cannot be the basis of an ethics of fraternity and peaceful coexistence among peoples and states." Three years later, at a Vatican conference, he said that "the threat of their use as well as their very possession is to be firmly condemned."

Protestant ethicist Paul Ramsey asked the following. Imagine that traffic accidents in a particular city had suddenly been reduced to zero, after which it was found that everyone had been required to strap a newborn infant to the front and back bumper of every car. Would this "solution" be ethical? He concludes that "this would be no way to regulate traffic even if it succeeds perfectly, since such a system makes innocent human lives the direct object of attack and uses them as a mere means of restraining the drivers of automobiles."[51] Ramsey's point, asserted by most ethical and legal doctrine, is that moral error lies first in the intention to do wrong and only later in the act itself. This is why intended wrong (such as homicide) is considered more serious than accidental wrong (such as manslaughter). Imagine society decreeing that in the event of murder, punishment would befall not only the murderer, but all his family and friends

---

[xv] The nuclear firepower assembled at this base is such that if the state of Washington were to secede from the United States, it would become the third largest nuclear weapons country.

as well.[xvi] And not since early medieval times has it been considered acceptable to prevent war by threatening the mass killing of innocent hostages. And of course, never in human history has that potential mass killing encompassed so many.

Moreover, what of the morality of holding hostage the billions of people whose governments *do not* use nuclear deterrence and who have neither voice nor vote when it comes to their fate—in other words, those additional hostages, beyond the citizens of the nuclear powers, who also find themselves tied to a bumper? The same can be asked about those innumerable other creatures in addition to humans—from hickory trees to halibuts, herons to hippos—whose future is in the hands of those few and fallible people who control the dials of deterrence. Supporters of deterrence claim that sometimes it is necessary to commit an evil (threatening nuclear war) to prevent a greater one (an actual nuclear war, or domination of one's country by a foreign power). On the other hand sits, as we have seen, the paradox that only by "meaning" such threats—deploying weapons that just might be used—can they be effective.

The United Methodist Council of Bishops went even further than their Catholic counterparts and refused to condone the threat implied by nuclear deterrence, regardless of its expressed intent.

> The moral case for nuclear deterrence [the bishops concluded], even as an interim ethic, has been undermined. . . . Deterrence no longer serves, if it ever did, as an acceptable strategy . . . . Deterrence must no longer receive the churches' blessing, even as a temporary warrant for the maintenance of nuclear weapons.[52]

HOW MUCH IS ENOUGH?

Let's continue our peek behind the many masks of deterrence by looking at this seemingly simple question: how much is enough? If it is acknowledged that we are sitting on a nuclear powder keg, then it is surprising how little thought has been given to how much powder should be in that keg.

Some people make what they see as a serious case for keeping firearms to protect their homes and themselves, but usually not a machine gun, bazooka, or flamethrower. Moreover, not in the dozens, hundreds, or thousands. What about nuclear weapons? There is no way for civilian or military leaders to know when their country

---

[xvi] And yet, Ramsey concluded that nuclear deterrence is nonetheless ethically acceptable, because, in this case, the end justifies the means—assuming, of course, that "the end" doesn't literally mean what it says!

has accumulated enough nuclear firepower to satisfy the requirement of having an "effective deterrent." The sky's the limit (not to mention outer space). For example, if one side is willing to be annihilated in a counterattack, it simply cannot be deterred, no matter the threatened retaliation. Similarly, if one side is convinced of the other's implacable hostility, or of its presumed indifference to loss of life among its own population, no amount of weaponry will ever seem to be enough. As long as money is made by accumulating weapons and as long as prestige, careers, and production jobs are served by designing, producing, and deploying new generations of nuclear hardware, there will be insistence on yet more of it. No country's gun closet, no matter how accommodating, will ever be adequate.

Adding to this dynamic is a primitive prenuclear psychology demanding that security requires having more weapons, whatever they are and no matter how many, than the other side.[53] This was an especially potent driver of nuclear accumulation during the overt Cold War arms race between the United States and the USSR. Here is former US Secretary of Defense, William J. Perry: "I can testify that during the Cold War, no US president was willing to accept nuclear forces smaller than those of the Soviet Union. And I believe that this perceived imperative did more to drive the nuclear arms race than did the need for deterrence."[54] Unlike a horse race or foot race, an arms race has no finish line, thereby guaranteeing no limit to the question: how much is enough?

And that's not all. Insofar as nuclear weapons also serve symbolic, psychological needs, by demonstrating the scientific and technological accomplishments of a nation, and thus conveying legitimacy to otherwise insecure leaders and countries, then once again, there is no rational way to determine the minimum (or cap the maximum) size of one's arsenal. At some point, additional detonations would come up against the law of diminishing returns or, as Winston Churchill pointed out, they simply "make the rubble bounce." Under the direction of Defense Secretary Robert McNamara, strategic planners in the United States tried to rationalize this problem, declaring during the 1960s that the inflection point—at which more explosive force generated relatively little effective destruction—was reached when roughly one quarter of the USSR's population was destroyed, along with one half of its industrial capacity. By a lucky coincidence, this could be achieved easily by obliterating the two hundred largest Soviet cities, which comprised one third of the country's population—a bit more than one quarter, but what's a few million additional deaths among enemies?

It was then proclaimed, for no objectively demonstrable reason, that this was the threshold for deterring the Kremlin. Just to be "safe," it was further decreed that *each leg* of the US strategic triad (bombers, land-based ICBMs, and submarine-launched missiles) should be outfitted with sufficient weaponry to do this, even without the other two.[55] Earlier, and even afterward, strategic planners on the Air Force and Navy staffs (less so the Army and Marines) engaged in an ever-upward-ratcheting cycle whereby identifying potential targets led to necessitating more bombs and missiles, which in turn exerted pressure to justify their existence by finding more targets.

During the Cold War, diplomat and historian George F. Kennan assessed the US–Soviet nuclear arms race as follows:

> We have gone on piling weapon upon weapon, missile upon missile, new levels of destructiveness upon old ones. We have done this helplessly, almost involuntarily: like the victims of some sort of hypnotism, like men in a dream, like lemmings heading for the sea, like the children of Hamlin marching blindly along behind their Pied Piper. And the result is that today we have achieved . . . in the creation of these devices and their means of delivery, levels of redundancy of such grotesque dimensions as to defy rational understanding.[56]

The grotesque dimensions of nuclear redundancy are not actually so much a result of irrationality as of a kind of rational understanding—the peculiar logic of deterrence—gone amuck. Imelda Marcos, wife of Philippine dictator Ferdinand Marcos, was obsessed with shoes, accumulating more than twelve hundred pairs. Deterrence has similarly made the United States and Russia no less obsessed with accumulating nuclear bombs and warheads, ostensibly to make their deterrence-based mutual threats more robust.

It has been said that compared to conventional weaponry, nuclear is a bargain, offering "more bang for the buck." They are nevertheless expensive. According to the nonpartisan Congressional Budget Office, current US plans call for spending roughly $1.7 trillion—a likely underestimate—on new nuclear weapons and their delivery systems over the next three decades. The price, however, can be deceptive. Earlier, our discussion of animal threats considered the role of costly signals, whereby honesty could be maintained and deception largely avoided because weak individuals could not masquerade as more formidable than they really are. Such handicaps do not work, however, in the world of nuclear threats. Thus, when China developed its nuclear arsenal—beginning during the 1950s and succeeding by 1964—it was economically weak and underdeveloped. Similarly, when he was president of Pakistan, Zulfikar Ali Bhutto announced that "even if we have to eat grass, we will make a nuclear bomb." And North Korea, with one of the world's weakest economies, was nonetheless able to build a nuclear arsenal, complete with missilized delivery systems.

The US government's Nuclear Posture Review is a strategic planning document that is revised every four years. The Trump Administration's version, issued in February 2018, was especially destabilizing. Whereas the Obama era review envisioned a reduction in the size and strategic importance of nuclear weapons, the Trump document called for precisely the opposite: expansion and "modernization." It also lowered the bar for their use, stating that they would be employed in response to a range of *nonnu-clear* provocations, including a cyberattack. It isn't clear whether this particular threat would be carried out or is intended for deterrent purposes only.

In 2016, the United States deployed 1,930 warheads, and had 4,670 stockpiled with 2,300 in retirement. Russia deployed 1,790, stockpiled 4,490, and had 2,800 in retirement. As of 2019, the active US nuclear triad was distributed as follows (not counting warheads and missiles in reserve), with the changes proposed by the Trump Administration in italics:

- Four hundred silo-based Minuteman III ICBMs, each carrying one warhead. *Six hundred new ICBMs—with four hundred to be actively deployed, and the remaining two hundred for spares and testing, costing more than $140 billion.*
- Up to 280 multiple-warhead Trident II D5 submarine-launched ballistic missiles carried aboard twelve *Ohio*-class submarines, each capable of firing twenty missiles (two additional subs and their missiles are usually out of commission at any given time for repairs and modernization); *twelve new* Columbia-*class subs, with new ballistic missiles and warheads, costing more than $128 billion.*
- Twenty B-2 stealth bombers, each capable of carrying sixteen gravity bombs, plus up to forty-six B-52H bombers, each capable of delivering twenty nuclear-armed, air-launched cruise missiles; *one hundred enhanced stealth-equipped B-21 Raider bombers, costing more than $97 billion.*[57]

Current nuclear numbers (even without the proposed increases), defy rational understanding, as Kennan pointed out four decades ago. Here is a story, beloved of psychiatrists: a patient complains that her husband thinks he is a chicken. The therapist suggests that the husband could benefit from treatment, to which the wife replies, "That's fine, but we need the eggs." It is downright crazy, defying rational understanding, but we need our nuclear eggs.

Donald Trump's nuclear shopping spree looks as though the Cold War never ended, and is liable to ensure that it comes roaring back. Between the fall of the Soviet Union in 1991 and the rise of new tensions since 2010, both the United States and Russia substantially reduced their nuclear stockpiles, going from a high of around seventy thousand to approximately sixteen thousand. So far, so good. However, goodness is not the whole story, because (1) both countries are developing smaller, more useable weapons that blur the boundary between conventional and nuclear weapons, and (2) the rise of irrational nationalism has led to widespread denial of science (and disparagement of genuine expertise), opening the door to reincarnated belief in a winnable nuclear war as well as fallacious revisiting of the supposed benefits of "strategic superiority."[58]

PRISONER'S DILEMMA AND FRIED CHICKENS

Deterrence is serious stuff. Not a game. Yet, it has been a favorite arena for practitioners of game theory—a logical system that in the world of strategic planning and analysis is typically cloaked behind screens of seemingly impenetrable jargon and symbols, yielding analyses that appear, or are made to appear, more complicated than they actually are. The effect is amplified when combined with top-secret information and computer-based decision-making algorithms. Many people are afraid of game theory because it can carry a heavy freight of mathematics. In fact, we all have good reason to be frightened by it, not because of the math, but because of how game theory has been used by strategists seeking to bolster deterrence. Paraphrasing Einstein, God does not play dice with the universe . . . but nuclear strategists do. The rarified domain of mathematically fancy nuclear threat-making and targeting scenarios risks carrying strategic analysts into their own fantasy world. But reality has a habit of persevering. It is, as Philip K. Dick once noted, "that which, when you stop believing in it, doesn't go away."[59]

In his influential book, *Thinking About the Unthinkable*, nuclear strategist Herman Kahn urged Americans to consider what he labeled a "post-attack reality":

> the possibility—both menacing and perversely comforting—that even if 300 million people were killed in a nuclear war, there would still be more than 4 billion left alive. . . . And a power that attains significant strategic superiority is likely to survive the war, perhaps even "win" by extending its hegemony—at least for a time—over much of the world. Reconstruction will begin, life will continue . . . .[60]

Many different kinds of ostensibly realistic games have been identified and studied, of which the two most important are Prisoner's Dilemma and Chicken.[61]

First, Prisoner's Dilemma, excluding the math. Imagine two prisoners, arrested for participating in the same crime. They both proclaim their innocence. The prosecutor can only get a conviction if they both plead guilty, so he manipulates them as follows, presenting each, independently, with the same fiendishly difficult dilemma.

"If you plead guilty, you'll spend a limited time in jail, perhaps a few years. If you both persist in pleading not guilty, I admit that you'll both do better yet—just a few months for resisting arrest—because I won't be able to convict either of you of any serious crime. So it might seem that cooperating with each other and pleading not guilty would be a good deal, and it could be, except for this: *both* of you would have to make the same not guilty plea. If you try to work together like that but at the last minute you plead guilty, I'll reward you by letting you go free (no jail time, the best payoff of all) and the evidence you'll provide against him will be enough to put him away for decades. And by the way, I'm making the same offer to him."

The choice faced by each prisoner might seem easy: cooperate with one another, both plead not guilty, and both get off with a light sentence. But it's not that simple, hence the dilemma. Here is a peek into the strictly logical thinking of either prisoner; because the situation is symmetrical, it applies to both.

"Regardless of what he promises me, I don't know what my accomplice will do, but he has two choices: plead guilty or not guilty. What's my best move? If he pleads not guilty, then I could take advantage of his not guilty plea and go for guilty, which would set me free while he rots in jail. On the other hand, what's my best move if he pleads guilty? In that case, I'd better plead guilty too, or else I end up the sucker who sits in jail while he goes free. Either way, therefore, my choice is clear: *Guilty, your honor!*"

What has happened here is that our prisoner is stuck, tempted by the opportunity to go free while his buddy is left a sucker. Simultaneously, he is fearful of being suckered himself. So he has been manipulated by the situation into pleading guilty, regardless of what the other fellow is going to do—that is, he finds himself forced to be uncooperative with the other player. This applies equally to both prisoners. The dilemma, therefore, is that by following unimpeachable logic of maximizing personal payoff regardless of what the other guy does, both players end up pleading guilty and as a result, serve several years in jail, whereas if they had cooperated they would have gotten off with just a slap on the wrist.

If the relevance of Prisoner's Dilemma to the world of threats—nuclear and otherwise—isn't clear, consider the situation of two countries trying to decide whether to abide by an arms control treaty. Each would be well served by doing so, because it would be less threatened by the other's weapons and would also be able to spend resources on pressing domestic needs. At the same time, each might well be tempted to cheat by developing and deploying weapons believed to give it an advantage over the other. In addition, each could also fear that the other might do precisely this, giving *it* an unacceptable advantage. Insofar as Prisoner's Dilemma provides a model for such situations, it suggests that each side would end up cheating, to their mutual disadvantage.

Renowned Japanese film director Akira Kurosawa portrayed an especially lethal Prisoner's Dilemma in his film, *Ran*, which was based on Shakespeare's *King Lear*—but much bloodier. Two rivalrous brothers, each leading an army, have called a battlefield truce, but it doesn't last long, because each fears that the other will conduct a devastating sneak attack. As a result, they blunder into mutual carnage. In this case, the "prisoners" each have two options: strike first or refrain. Out of anxiety that the other will break the truce and thereby gain an advantage—while also tempted to gain that same advantage for himself—they both end up striking "first," with an outcome as tragic as it is predictable.

Of course, Prisoner's Dilemma is greatly oversimplified, and thus not very realistic. Countries, like people, normally have options beyond just cooperate or defect; they can also communicate and thus provide reassurance. (In the strict Prisoner's Dilemma model, the two participants are isolated, neither one having any idea what the other is up to.) In addition, most participants in such interactions—unlike the hypothetical prisoners—want not just a single interaction; rather, they anticipate a series of such exchanges, resulting in what theorists call an "iterated game." Prisoner's Dilemma is nonetheless beloved of strategists because it provides a useful model that can then be tested against various tactical moves, and also because it helps justify their own coldly logical, often amoral, and uncooperative predispositions.

Although superficially similar to Prisoner's Dilemma, the game of Chicken is a different bird altogether. It was initially presented to the American public as a daredevil car stunt, immortalized in the 1955 James Dean movie *Rebel Without a Cause*. Two amped-up young men drive toward a cliff at high speed. The first to bail out loses; the one who perseveres is the winner. The goal in a game of Chicken is to induce your opponent to reveal his cowardice by jumping out of the car first, while you prove your manhood by staying in until the last possible moment.

Philosopher and antinuclear campaigner Bertrand Russell made an even more dramatic analogy between strategic games of Chicken and two drivers heading toward each other, each straddling the white line, seeking to induce the other to turn aside. The most noteworthy example of nuclear Chicken occurred during the Cuban Missile Crisis. Let's briefly revisit this event.

The United States demanded that the missiles be withdrawn. The Soviets refused. After considering and rejecting various options—including a conventional attack on the missile sites, an invasion of Cuba, and a preemptive nuclear strike against the Soviet Union—President Kennedy decided on a naval blockade, designated at the time a "quarantine" because, as mentioned earlier, under international law, a blockade is an act of war. JFK subsequently stated he thought the chance of nuclear war had been "between one in two and one in three."

As then-Secretary of State Dean Rusk put it, "We were eyeball to eyeball, and the other guy blinked." In other words, the Soviets turned aside in that game of nuclear Chicken. It is also possible that Premier Khrushchev decided that keeping nuclear missiles in Cuba was not a big deal after all, certainly not worth risking the lives of millions. In any event, the Soviets quickly responded to that humiliation by massively expanding their nuclear forces. After all, one way to win a game of Chicken is to drive a big, threatening, armor-plated vehicle, so that in the event of a collision, the cost to your opponent would exceed the cost to yourself. Even someone so stupid as to play Chicken in the first place would doubtless swerve if his small sports car were up against a massive, armored cement truck.

Having swerved in Cuba, the USSR substantially increased the size of its vehicle—as did the United States—so that the US and Russia have since been equally capable of destroying the other. As a result, in a repeat confrontation, neither side is likely to accept the ignominy of being the "swerver." Americans tend to think of the Cuban Missile Crisis—when they think of it at all—as a triumph for the United States and even, somehow, for deterrence, because the Soviets swerved. Missing here are three important realities. First, deterrence failed insofar as President Kennedy was definitely not deterred by the threat of a Soviet nuclear response; worried, yes, but deterred, no. Second, a consequence of that failure of deterrence is that Kennedy was willing to trust the fate of Earth to a decision by Khrushchev and the old men of the Politburo. In other words, the US President surrendered control over the most important of all decisions.

And finally, the outcome actually had dire consequences because it propelled the USSR to mount a crash program to increase its nuclear arsenal, leading to a situation of parity—which might sound desirable—except that when playing Chicken, it can be lethal. When each driver insists that the other turn aside, with the cold logic of deterrence trumped by anger, stubbornness, pride, worry about loss of face (or elections, as was a contributing factor in JFK's determination to plow straight ahead), the result is liable to be fried chicken.

Undeterred himself, nuclear strategist Herman Kahn made the provocative suggestion that when two vehicles are evenly matched, another way to win the game of Chicken is to throw your steering wheel out the window. This would demonstrate that, although swerving would be rational, you can't do it, so the other driver must! The winner would then be the one who discards his steering wheel first, making it a contest to see who is quickest to disavow control, leaving the other obliged to swerve—assuming, of course, that the other driver is rational.

Something analogous to tossing the steering wheel has been seriously proposed: given the massive problem of making nuclear threats credible, human beings should be taken out of the loop, with launch authority delegated to autonomously functioning satellite sensors and computer algorithms. The resulting arrangement has been called "launch on warning"—a system intended to make sure that the other swerves by announcing that, although retaliation would be irrational and even self-destructive, it will happen nonetheless when radar and satellite sensors detect an incoming attack. It essentially says we have already discarded our steering wheel, so don't think that we might swerve, because we can't. At the same time, given the woeful record of false alarms and computer malfunctions, this effort to bolster deterrence carries the risk of bringing about precisely the result it would be designed to deter.

It appears that automatic, computer-controlled launch on warning has not been instituted in the United States, despite a growing concern among some nuclear hawks that counterforce targeting has already made it necessary. Not so the Soviets. After the end of the Cold War, it was revealed that the USSR had deployed its "Dead Hand" system, which might be called "fail-dead" (as distinct from "fail-safe"), designed to retaliate lethally against NATO if the Soviet motherland and its leaders were obliterated in a first strike. To the amazement of all, it turned out that—just as in the movie *Dr. Strangelove, or How I Learned to Stop Worrying and Love the Bomb*—the USSR's top brass had never informed the United States about Dead Hand.

Keeping its existence secret, in a cinematic Cold War satire as well as in reality, was equivalent to jettisoning the steering wheel, but keeping *that* secret. It brings about the worst of both worlds, depriving the Dead Hander of the opportunity to second-guess a malfunction, while doing nothing to inhibit an opponent. Even something as potentially lethal as Dead Hand or launch on warning, bad as it is, can't bolster deterrence if the other side doesn't know that it exists. If nothing else, this omission by Soviet leadership should end any illusion that those in charge of deterrence are smarter than the rest of us and therefore can be trusted to deal appropriately with their immense responsibilities. It isn't known whether Russia has retained, modified, or abandoned Dead Hand.[xvii]

There are other ways, at least in theory, to win a game of Chicken. You could show up drunk or act like a raving lunatic—someone who is "just so crazy he might do it," with "it" being refusing to swerve, no matter what. Another way to get the other guy to back down is to threaten that you don't really care about a potential collision, either because you are so committed to winning or because, for you, the cost of swerving is too great. Or because you're a bit nuts. During the Vietnam War, President Nixon told his chief of staff, H.R. Haldeman, that he wanted to try out the "crazy man theory," whereby Ho Chi Minh was to be told the US president was increasingly irrational and, therefore liable to use nuclear weapons against North Vietnam. So, the North had better accede to US demands. It didn't work.

Donald Trump is not known for his academic skills or historical acumen, so it is unlikely that he read Kahn's proposals.[xviii] But, his combination of being unpredictable, unreliable, disconnected from reality, possibly unhinged, and definitely obsessed with

[xvii] This curious dynamic brings to mind the "beheading game," most famously developed in the medieval romance *Sir Gawain and the Green Knight*. A huge, heavily armed, and mysterious green knight appears at King Arthur's court, challenging the king to strike him however he chooses with a sword, after which he must then meet the green knight a year later, to be similarly assaulted in turn. Sir Gawain, substituting for Arthur, promptly cuts off the green knight's head, which is then picked up and reaffixed by the stranger, who announces that he looks forward to giving Gawain his comeuppance next year. In this chivalric romance, Gawain avoids his seemingly lethal fate, but in its manifestation via nuclear deterrence, the beheading game is real.

[xviii] Since the dawn of the nuclear age, every US president but one has acknowledged being humbled when confronted with nuclear realities. During his presidential campaign and while in

winning is a conjunction that Herman Kahn might have applauded. But not many others.

The biological world offers a revealing parallel. Adult male Asian elephants occasionally go into "musth," a peculiar state in which a foul-smelling material oozes from their eyes, often affecting their breath and urine as well, and they behave more than a bit crazy: highly aggressive, unpredictable, and unresponsive to the usual niceties of elephant sociability. This is generally a time-limited condition during which these individuals—sometimes known to Westerners as "rogue" elephants—are especially dangerous. Not just people, but other elephants accordingly give them a wide berth. They swerve, understanding that such an animal is not to be trusted to play by the rules—that is, he won't turn aside, so they had better do so.

In the game of Chicken, it is also possible that if a driver believes he is unlikely to be seriously injured in a crash—or if the other driver thinks that his opponent believes this—then the one with a sense of his invulnerability will be less likely to swerve, while the other one, anticipating this development (regardless of whether a survivorship asymmetry is genuine) will be more inclined to do so. During the Kennedy Administration, blast and fallout shelters were encouraged in the expectation that "civil defense" would help the United States come out on top, not only in the event that deterrence failed, but that it would also make the Soviet Union more disposed to take American resolve seriously.

This view was resurrected during the early Reagan years. Especially notorious was the claim by T.K. Jones, Deputy Undersecretary of Defense for Strategic and Theater Nuclear Forces, who argued for the survivability of nuclear war, urging that all that was needed was for Americans to dig a trench, cover it with a sturdy wooden door, pile dirt on top, and crawl underneath. "If there are enough shovels to go around," Mr. Jones announced, "everybody's going to make it. It's the dirt that does it."[62] (In response, an eighty-year-old-woman wrote to her local newspaper that she was eager to do her part for national defense, but that her arthritis was severe and, moreover, she didn't understand how she was going to heap dirt upon the door and *then* crawl underneath!)

Yet another possibility could arise if there is a different kind of asymmetry between the players. If, as in the iconic, movie-based version, both of the drivers are eager to impress a girl (or anyone else), but one is known to be madly in love whereas the other is less committed, the likelihood is that the former would be more likely to run risks to win, and the latter less so. The love-besotted driver therefore has an advantage in that the cooler one would be more likely to swerve, knowing that the other's obsession

office, Donald Trump bragged several times about being an incomparable genius, with a sophisticated understanding of nuclear weapons, based on his uncle, John Trump, who had been an engineering professor at MIT: "We used to talk a lot about nuclear."

might even lead to willingness to die rather than lose. This disparity may converge with observations of animal conflict, whereby territory owners typically come out ahead in mutual threat contests, perhaps because they are more invested in the property, den, nest, mate, or whatever else is being contested.

All of this has implications in the world of international conflict and crisis. When it is clear that one side has a serious strategic interest in a particular outcome whereas the other's national security is less engaged, and if the two sides are otherwise evenly matched, the one who cares less nearly always defers to the one who cares more. For example, the United States did not respond militarily when, in 1956, the Soviet Union crushed the brief freedom movement in Hungary. At the time, Hungary was one of the USSR's Eastern European satellite countries, perceived as a strategic buffer against NATO and the West, whereas the West—although sympathetic to the plight of Hungarian freedom fighters—kept its distance from a dispute that was, frankly, comparatively distant.

Six years later, when the Cuban Missile Crisis emerged, things were reversed: the United States was deeply agitated by the prospect of nuclear weapons just ninety miles from its shore, whereas for the Soviets, backing down wasn't nearly as painful as it would have been for the Americans. In addition, President Kennedy believed at the time that a failure to respond forcefully would cost his party the forthcoming off-year elections and might even result in his impeachment. The fact that his political survival may well have depended on his success doubtless stiffened JFK's spine and made him unlikely to swerve—something that Premier Khrushchev also knew because the president's brother, Robert Kennedy, had conveyed this information to Anatoly Dobrynin, the Soviet ambassador in Washington.

Having a commitment is closely related to the question of credibility, which we visited earlier: insofar as one's opponent knows about the commitment, the committed side has greater credibility. And this, in turn, bleeds into the deterrence-relevant question of saving face, because both sides are liable to be especially concerned about their reputations—notably, that they are known to react firmly and fiercely as needed. Once again, Thomas Schelling had an important and disturbing insight. He argued that, particularly in a world of deterrence, face—and one's reputation of being obsessed with maintaining it—is "one of the few things worth fighting over." Schelling suggested that even though "few parts of the world are intrinsically worth the risk of serious war by themselves, especially when taken slice by slice," sometimes commitments are nonetheless worth making. This holds, he claimed, in order to maintain "a country's reputation for action, the expectation other countries have about its behavior,"[63] and even if the action in question might involve massive loss of life.

Deterrence-based pressure to cultivate an image of stubborn recklessness can therefore push a country's leaders to engage in lethal actions for no reason other than to prove their willingness to do so. According to one highly regarded strategist, "In order to buttress its credibility, a nation should intervene in the least significant, the least compelling, and the least rewarding cases, and its reaction should be *disproportionate* [emphasis added] to the immediate provocation or the particular interest at stake."[64] This prospect is not as outlandish as one might expect.

For example, Japan and China both claim ownership of some uninhabited, rocky, and altogether valueless islands known in Japan as the Senkaku Islands and in China as the Diaoyus. They are currently administered by Japan; but what if China seizes one of them? Should the United States, given its extended deterrence treaty with Japan, respond militarily? With nuclear weapons? Thereby reacting in a disproportionate way to this exceptionally uncompelling situation in order to maintain the credibility of its commitment to extended deterrence? Only in the through-the-looking-glass world of nuclear deterrence can serious adults conclude, with a straight face, that it is worthwhile to precipitate massive carnage over issues that are not only insignificant, but to do so *because* they are insignificant.

The two crucial games, Prisoner's Dilemma and Chicken, are similar in their structure, but also crucially different. They are similar in that each involves just two players, both of whom have just two options: cooperate or defect in Prisoner's Dilemma, which is equivalent to swerving or going straight in Chicken. They also share a key characteristic in that both games feature the highest payoff for the player who is nasty at the same time that the other is nice, with the next highest payoff going to both if they engage in mutual, nice cooperation (Prisoner's Dilemma) compared to mutual swerving (Chicken). But then the two diverge significantly. The worst payoff in Prisoner's Dilemma goes to the participant who tries to cooperate and finds herself paired with a selfish and uncooperative defector, whereas mutual defection—the mathematically and logically stable outcome—results in the next-worst result. When it comes to Chicken, these two prospects are reversed: it is undesirable to be the one who swerves, thereby being labeled a chicken and losing the prize if at the same time the other driver wins with the tactic, "Damn the torpedoes; full speed ahead!" But, unlike Prisoner's Dilemma, the worst outcome in Chicken, and one shared by both players, arises if neither swerves and a crash ensues.

Optimal strategies therefore differ for the two games. When playing Chicken, the idea is to send signals of toughness or invulnerability—of resolve, lunacy, indifference to a crash, and so forth—seeking to persuade the other player that you are not going to swerve (or, in the steering-wheel example, that you have made yourself unable to do so) no matter what. In short: increase the threat you pose. At the same time, Chicken requires a careful titration of that threat, making it sufficiently believable and serious to ensure it is effective, but not so imminently terrifying as to cause the other side to conclude that, because all is lost, they might as well keep on—possibly right into your oncoming vehicle.

Of course, games of Chicken aren't limited to their nuclear manifestation. As of mid 2019, the Trump Administration had unilaterally withdrawn from a nuclear agreement with Iran that had been negotiated by the Obama Administration, and in which Iran had been in full compliance. The United States resumed severe economic sanctions on the Iranian economy, which had previously been loosened as part of the deal whereby Iran accepted restrictions on its nascent nuclear weapons program. The Iranian government, pressured by this and also not wanting to be (or to seem) intimidated, appears to have responded by increasing support for violent extremists in its region as well as sponsoring limited attacks on Persian Gulf shipping. It also began uranium enrichment beyond what had been prohibited by the earlier agreement. The United States, in turn, increased military deployments in the Gulf. Both sides, in short, sent messages of toughness and resolve, maximum threats and minimum diplomacy, directed toward getting the other to swerve.

By contrast, when playing Prisoner's Dilemma, smart signaling involves sending messages of reduced threat, of reassurance and conciliation, trying to persuade the other player that she needn't worry, that you are going to cooperate and be nice, and are thus a suitable partner, willing to settle for the moderate but positive shared outcome of mutual restraint. A problem, however, is that having persuaded the opponent of her benevolent intent, a purely self-interested Prisoner's Dilemma player could always defect at the last minute, thereby reaping the payoff of doing so and leaving the other one a sucker. And, of course, that side, knowing this, and also being similarly tempted, might be all the more liable to do the same thing.

When people engage in simulated games of Prisoner's Dilemma, there is a tendency for them to cooperate more than a strictly logical calculus of game theory modeling would predict. So maybe there is hope of escaping from this particular dilemma after all. Interestingly, however, when such simulations are run using experts in game theory—typically economists at a professional conference—defection (that is, noncooperation) is far more common.[65] Economics really does seem to warrant its moniker as the dismal science. All the more reason, perhaps, to keep certain kinds of experts away from competitive-versus-cooperative decision making.

In the movie *War Games*, a young computer hacker almost starts World War III when he runs a nuclear war simulation that a supercomputer mistakenly interprets as the real thing. At the end, with annihilation narrowly averted, the computer notes that the whole business is "A strange game," before concluding, "The only winning move is not to play." In an especially nice touch, the movie ends with the computer suggesting an alternative: "How about a nice game of chess?"

STRATEGIC DEFENSE?

Even some cheerleaders for nuclear weapons have shown questionable enthusiasm for deterrence by supporting various substitutes. Of these alternative schemes, the most popular—and the one that has generated the most expense and controversy, and that has caused the most additional international tension—has been the attempt to develop and deploy what President Reagan called a "Strategic Defense Initiative," and what critics (following the lead of Senator Edward Kennedy) dubbed "Star Wars." The idea sounds good: an array of radar-guided missiles would defend the United States against a nuclear attack. Easier said than done, however. A feat described as hitting a bullet with a bullet, in practice it has been bedeviled by a huge number of technical problems, including the very short time frame in which it must operate, the relative ease with which such a system could be overwhelmed by multiple warheads, or its radar confused by metallic chaff.

Moreover, insofar as a defending side seeks to respond to offensive tactics intended to overwhelm it, this leads, in turn, to yet more offensive tricks, and so on, ad infinitum. In short, any antiballistic missile system would be vulnerable to the "fallacy of the last move," and would also have to be fool-proof to be effective. If just one incoming warhead makes it to just one city, the devastation would be immense. The record thus far for successful interception of incoming dummy warheads by what is now officially labeled "ballistic missile defense" has not been encouraging, even when the tests have been highly scripted, with the exact time of the launches known in advance, along with the anticipated trajectory of the "incoming."

Even less encouraging is the effect that strategic defense—or merely the prospect of it—threatens to have on the already shaky stability of deterrence itself. Recall that deterrence stability (the Holy Grail of most analysts) requires a balance of terror whereby each side is inhibited from attacking by the certainty that doing so would bring about unacceptable retaliation. Moreover, not only is a first-strike attack supposed to be deterred as a result, but the participants must also be assured that, because deterrence will inhibit any nuclear adventurism, no one is tempted to preempt such an attack by an opponent, even under crisis conditions.

Let's imagine, however, that—contrary to the laws of physics and mathematics— one side succeeds in deploying a ballistic missile defense. Having done so, the prospect could then beckon for the deploying country to attack the other side, destroying a substantial number of the victim's nuclear weapons and delivery systems. The overwhelming likelihood is that some would nonetheless survive and be capable of retaliating—except that if the attacker had deployed even a less-than-perfect ballistic missile defense, it might be able to shoot down the "ragged retaliation" that would be the best that a victim could manage. And this anxiety, in turn, could start a feedback process of unending instability: whoever possesses a missile defense system might feel emboldened to attack because of having it, while at the same time, a potential victim, fearing such an attack, might feel desperate and threatened enough to initiate a preemptive strike, which could lead the country with the missile defense system to worry that this might be in the works and to preempt the preemption—and so on. This dangerous dynamic, which we already confronted with respect to counterforce, is exacerbated by ballistic missile defense.

Bear in mind that strategic analysts are enamored of worst-case scenarios,[xix] which demands that they consider this sequence seriously. The Soviets—and now, the Russians—have long had exaggerated respect for American technological prowess (understandable, given that the United States has led nearly every dimension of their nuclear competition) and, as a result, have strenuously opposed antiballistic missile deployment. Ronald Reagan, who evidently was quite unsophisticated when it came to science, and who may have sincerely believed that missile defense was both feasible and nonprovocative, claimed in his first Star Wars speech (1983) that such a defense system would supersede deterrence, making nuclear weapons "impotent and obsolete."

But deterrence was—and still is—the major justification for these weapons, and even the possibility of downgrading it evoked shudders from the pronuclear establishment. Accordingly, in Reagan's next speech on the same topic, he backpedaled, claiming that Star Wars would "enhance deterrence."[66] Soviet officials were shuddering too, worried that Star Wars would make *their* weapons impotent and obsolete, leaving the United States with a dangerous and destabilizing advantage. At a summit meeting in Reykjavik, Iceland, in 1986, Reagan and Mikhail Gorbachev almost reached an agreement to eliminate all nuclear weapons. The sticking point was Reagan's insistence on pursuing strategic missile defense and Gorbachev's equal insistence that it be prohibited.

The issue is alive today. Under the George W. Bush Administration, the United States unilaterally withdrew from an antiballistic missile treaty that had been signed in 1972, in which the two countries committed themselves to mutual restraint in the service of deterrence. The United States periodically resuscitates plans to implant such a system in Eastern Europe, officially to intercept a possible limited attack from rogue Russia, but reviving Russian fears of a more sinister purpose. China's concern is more justified than Russia's, because the Chinese have opted for "minimal deterrence," having deployed only about 280 to 300 nuclear warheads, enough by their analysis to inhibit a full-on attack by either Russia or the United States. In a rational world, Russia ought to be able to rest easy that, even after suffering a first strike and even in the unlikely

---

[xix] Except when it comes to taking seriously the worst-case view that deterrence will likely fail over time.

event that the United States was so foolish as to rely on an antiballistic missile astro-dome from sea to shining sea, thereby rendering a counterattack unavailing, some of their retaliatory missiles would doubtless get through, which in turn should prevent any initial attack.

By contrast, China, with its much smaller nuclear arsenal, could have reason to fear a US antiballistic missile system, which threatens to obviate its relatively small nuclear missile force, even though the US missile defense systems are widely acknowledged to be less robust than advertised—except, of course, when such advertisement is made by military officers and civilian contractors testifying before congressional appropriation committees. The Chinese leadership has therefore been expressing loud and increas-ingly angry opposition to US deployment of even a rudimentary antiballistic missile system in South Korea, which theoretically poses a threat to their country's minimal deterrence, although it is ostensibly intended to deter North Korea.

The negative shadow cast by antiballistic missile systems goes further yet. Russian fear of such a US system motivated Vladimir Putin's government to invest heavily in developing weapons intended specifically to avoid any "defensive" antimissile deploy-ment by the United States, even if the latter is widely acknowledged to be unfeasible. Thus, in 2018, Putin announced the development of an autonomous undersea drone (subsequently labeled *Poseidon*), which—outfitted with a nuclear warhead—could slip under any Star Wars system. In addition, the Russians initiated research into what appears to be a nuclear-powered cruise missile that could, in theory, fly around—once again, beneath the interception capabilities of any defensive array—for days, weeks, even possibly months to retaliate against perpetrators of a first strike even after all Kremlin military and political leaders have been incinerated. In 2019, an early proto-type of such a weapon exploded at a Russian testing facility near the White Sea, killing a number of Russian scientists, but emphasizing the real-world consequences of both relying on deterrence and, paradoxically, undermining it.

As if this threatened revitalization of a post-Cold War arms race were not bad enough, the United States, Russia, and China have begun to develop hypersonic mis-siles, the extreme speeds of which would enable them to outwit any known defensive weaponry and also to attack more rapidly than ever, leaving victims less time than ever to respond. This is part of a new arms race in speed, exemplified by a worldwide push, led by the United States, to develop "prompt global strike" weapons. (Recall the security dilemma.)

It is difficult to imagine how anyone will be better off when such systems are in place—not just in the United States, but worldwide. And yet, the US is currently committed to a highly threatening and likely destabilizing development and acquisi-tion policy called "overmatch," which requires that it keep military superiority over Russia and China and everyone else, into the infinite future. Specifically, the Trump

Administration's National Security Strategy, released in February 2017, states, "The United States must retain military overmatch . . . so that it is capable of operating at sufficient scale and for ample duration to win across a range of scenarios." These include nuclear, conventional, unconventional, cyber, and space.

In the hilarious Charlie Chaplin film *The Great Dictator*, the Hitler and Mussolini characters sit in barber chairs, each trying to pump up his perch so as to be higher than the other. Overmatch cannot be mutual.

A key take-home message is that treaties—especially those involving nuclear weapons—may well deserve much of the credit for why we are still around, and thereby afforded the opportunity to debate their usefulness. This makes the Trump Administration's discard of these agreements particularly ill-advised. We don't have the luxury of a robust strategic safety net; but, as Michael Krepon, a long-time voice of nuclear sanity, pointed out in a *New York Times* article titled "The New Age of Nuclear Confrontation Will Not End Well," treaties are a big part of what exists:

> The thickest weight-bearing strands of this safety net are treaties that have been cast aside without being replaced. With diplomacy sidelined, policymakers are left with nuclear threats to deter competitors. But deterrence rests on the underlying possibility of the use of nuclear weapons; otherwise, they would cease to deter. This can lead to tragic miscalculations. Treaties, by contrast, muffle and reduce threats. So "strengthening" deterrence without treaties and diplomacy is dangerous; it's a recipe for threatening your way into tight corners . . . .[67]

HOW DO YOU SOLVE THE PROBLEM OF KOREA?

There is considerable debate among critics who are not immersed in the prodeterrence establishment: is deterrence the real intended goal of US—and Russian—nuclear policy or is it a smokescreen for a planned first strike? (Not surprisingly, hawks on each side argue that this is precisely what *the other* is up to.) When it comes to the problem of Korea, diagnosis is easier than solution, although the former leads to a reasonable approach to the latter. In any event, there should be little debate over the intentions of the North Korean regime.[xx]

[xx] North Korea's nuclearization also adds to the perennial discussion among strategic analysts regarding "intentions" and "capabilities" of a perceived opponent: although the intentions of most would-be nuclear proliferators may be debatable, there is typically much less uncertainty about capabilities. Moreover, the fact that North Korea—one of the world's most isolated and impoverished countries—was able to develop nuclear weapons and delivery systems makes it clear that *any* country could do so.

The Kim dynasty, of whom Kim Jong-un is the third manifestation, has long had one overriding goal:—staying in power. Although murderous and ruthless toward his own people, the current Kim and his cohorts are not suicidal. It is clear that their nuclear intentions are not to attack the United States, South Korea, or Japan, thereby ensuring their own destruction, but to deter—that is, to deter *us*.

It is hard to imagine that US officials, even the most hardline opponents of the Kim regime, are not aware of this. In fact, this might be precisely why the North's nuclear arsenal—and the possibility, someday, Iran's—is so galling, not so much a direct threat to the US homeland as a threat to the US military's freedom of action. A position paper by the Project for a New American Century, written in 2000 and titled "Rebuilding America's Defenses," argued that the United States

> must counteract the effects of the proliferation of ballistic missiles and weapons of mass destruction that may soon allow lesser states to deter U.S. military action. . . . In the post-Cold War era, America and its allies, rather than the Soviet Union, have become the primary objects of deterrence and it is states like Iraq, Iran and North Korea who most wish to develop deterrent capabilities.[68]

The concern of this "project," which was staffed by some of America's most influential hawks, is not so much that the United States will be attacked by these countries, but that unfriendly states may be able to prevent the US from acting against those it considers unfriendly.

To most Americans, it must seem ridiculous that the North Koreans should worry about a possible attack by the United States. After all, we're the good guys! Things look different, however, from Pyongyang. Go back, as their memories assuredly do, to the Korean War. There is little doubt that "they started it" by invading the South, but also little doubt that the devastation ultimately visited upon the North was overwhelming, leaving its people and its leadership with traumatic scars. At the same time, Americans, seeing the North as the aggressor, and aware of US casualties, are by and large unaware of the suffering that North Korea experienced at our hands.

The (conventional) bombing of the North by B-52s was long, leisurely, and merciless, even by the assessment of America's own leaders. "Over a period of three years or so, we killed off—what—20% of the population," Air Force General Curtis LeMay, head of the Strategic Air Command during the Korean War, told the Office of Air Force History in 1984.[69] In the twenty-first-century United States, that would be equivalent to killing sixty million Americans. Whether they deserved it is not the point; rather, what is relevant is the collective impact of the North Korean experience. Dean Rusk, a supporter of the war and later secretary of state, said the United States bombed "everything that moved in North Korea, every brick standing on top of another." [70] After running out of urban targets, US bombers destroyed hydroelectric and irrigation dams in the later stages of the war, flooding farmland and destroying crops.

Add to this the annual joint military exercises long conducted by South Korean and US forces, plus regular run-through practice sessions by commando units aimed at "decapitating" the North's leadership. Little wonder they—especially the upper levels of government and the military—are afraid of us. Yoon Young-kwan, a former South Korean foreign minister and professor emeritus at Seoul National University pointed out that even if it had not previously been bombed to smithereens by the United States, "A country like North Korea—a small and weak country diplomatically isolated and economically devastated and a country surrounded by big powers—may feel very insecure, even though its neighboring countries have no intention to attack them." [71]

In January 2018, Kim Jong-un reminded the world that a nuclear launch button is "always on my table" and that he could attack the continental United States with nuclear missiles: "This is a reality, not a threat." Of course it was a threat, and it still is. Not that North Korea would mount a first strike, but if the United States attacked him, it would regret it. To this, Donald Trump tweeted that his "Nuclear Button" was "much bigger & more powerful."

In 2000, President Kim Jong II (Kim Jong-un's father) told a South Korean newspaper that

[Our] missiles cannot reach the United States, and if I launched them the U.S. would fire a thousand missiles back and we would not survive. I know that very well. But I have to let them know I have missiles. I am making them because only then will the United States talk to me. [72]

And, he could have added, only then will the United States refrain from invading me. Shortly thereafter, when the US invaded Iraq, the leadership in Pyongyang concluded that Saddam had erred by not having his own nuclear weapons, whereupon the North escalated its nuclear program. By 2018, having established its nuclear bona fides, the current Kim can afford to tamp down his country's accumulation of missiles and warheads, confident that he will not meet the same fate as Saddam Hussein or Libya's Moammar Quadaffy.

Concurrently, Kim announced his *byungjin* (parallel advance) policy, which calls for stockpiling nuclear weapons in the belief that this would allow the country to focus on economic recovery. Since taking power in 2011, Kim has increasingly adopted *byungjin* as his national strategy, despite warnings from Washington and Seoul that it would further isolate his already heavily sanctioned country.

Given the outlandish bluster that long emanated from Pyongyang's leaders, it might seem odd that they fear us. After all, their threats—such as, to turn the United States into a "sea of fire"—were intended to make us fear *them*. All the better to deter us, my dear. But the North's clumsy efforts succeeded all too well—a case of deterrence run amuck. Maybe North Korea's arsenal indeed deterred the United States from attacking. More likely, however, if such an attack were to happen, it would ironically be because of their backfiring efforts to achieve security by threatening the US. Long before developing its modest nuclear arsenal, the North deployed thousands of conventional artillery pieces aimed primarily at Seoul, with its population of more than ten million, which is just thirty-five miles from the demilitarized zone and well within artillery range,

The reality is that its artillery provides the North with more than enough conventional firepower to deter any attack from South Korea or the United States. Having added nuclear weapons to its arsenal, the Kim regime has unnecessarily raised the ire and anxiety of its potential opponents and endangered itself and the entire Korean peninsula, plus the rest of the world. Because the North's nonnuclear capacity is well known and highly respected by both South Korean and US officials, its seemingly desperate efforts to add nuclear weapons to its armamentarium has counterproductively raised fears, exaggerated though they are, that the Kim regime actually has plans to attack the South, or even the United States. These "fears" have been particularly useful for such unreconstructed Cold Warriors as John Bolton (for a time, President Trump's National Security Adviser), as a convenient way to justify pressure against Pyongyang—and possibly even a military attack, ill-advised as that would be.

Even as some in the West have overestimated the threat posed by the North, the Kim regime likely underestimated Donald Trump's egotism and volatility, and ended up evoking his own threats of "fire and fury," along with serious talk of an impending "bloody nose" attack as punishment for their threats and also to show that we aren't intimidated. Shades of Chicken: when neither side will swerve, the outlook is grim.

Trump's statements toward North Korea (at least before the sudden and overhyped rapprochement of 2019) differed radically from long-standing US policy regarding deterrence, which had been that if anyone attacked the United States or any of its designated allies, they would be attacked in return. Rather, it was that "fire and fury" would result if they merely dared to "make any more threats." Clearly implied as well was that if North Korea developed the capacity to attack the US—already possessed by Russia, China, India, and, for what it's worth, the United Kingdom and France—it would be destroyed. Well, they have, and they weren't.

Admittedly, both before and during his presidency, Trump established a new level (qualitatively and quantitatively) of out-and-out lies, such that friends and foes all have good reason to disregard much of what he says . . . and nearly everything that he tweets.

But when it comes to nuclear threats, a unique dimension is engaged, and because careless talk can have extraordinary consequences, it can be extraordinarily dangerous.

Anxiety about war between the United States and North Korea was additionally ratcheted up by Trump's grotesque misunderstanding of the effectiveness of US ballistic missile defense. Thus, in October 2017, during an interview with Sean Hannity on Fox News, he raised the eyebrows of pretty much anyone who knew anything about nuclear weapons when he proclaimed, "We have missiles that can knock out a missile in the air ninety-seven percent of the time, and if you can send two of them, it's going to get knocked down." This claim is wildly out of line with even the most exaggerated overestimates made by the US Missile Defense Agency, which routinely hypes its own success rate to increase Congressional funding. The danger is that if the person with sole responsibility for ordering use of American nuclear weapons really believes his own lie, he might well think he could safely initiate an attack, thinking that US missiles would intercept the victim's retaliation.

They wouldn't.

Fortunately, at least as of 2020, both sides have calmed down. But it remains to be seen how long this breathing space will last. Trump claims that because he and Kim entered into their surprise bromance, North Korea is "no longer a nuclear threat." Not so. On the other hand, it never was the nuclear threat that Trump claimed; moreover, his exaggerated response—matching Kim's exaggerated deterrence-based swagger— initially made things hugely worse. As to whether Trump or Kim deserve credit for having tamped down tensions, imagine that someone entered your house and doused it with gasoline; would you then consider him a hero if he didn't actually light a match?

FAILURE *IS* AN OPTION

There once was a turkey who gobbled with glee,
"Oh fortunate, fortunate, fortunate me,
How lucky I am to be safe on this farm,
Where they feed me good food and prevent any harm."
Happy and confident, glad he was living
In such a safe place—on the eve of Thanksgiving.[xxi]

If nuclear weapons are ever used—not merely in threat, but in reality—deterrence will have failed. And yes, failure *is* an option. Should this happen, there will likely be no strategists or historians left to debate the reasons, or how deterrence fell short of its

[xxi] Written by the author.

promise to keep the peace. Reliance upon deterrence has involved its own persistent failure, intentional or not—a failure to look honestly at how it could fail. There may well be no other circumstance in which so many have risked so much while being so ignorant of what is at stake.

When people respond to threats, especially those that do not emanate from other people, their actions typically do not make the threat more likely to come about, or more destructive if it does. Getting a flu shot does not guarantee immunity, but it definitely reduces the likelihood that you will get sick and, moreover, makes it more likely that if immunity fails, the illness will be less intense than otherwise. Buying flood, fire, or car insurance doesn't change the probability that purchasers will suffer flood or fire, or get into a car crash, but they diminish the financial pain if things go badly. By contrast, the gravamen of this section has been that nuclear deterrence is qualitatively different. For one thing, because it encourages a country to accumulate nuclear weapons in the first place—not to mention often multiplying their number and lethality— it increases the actual damage to be wrought if and when deterrence fails. And for another, it risks making nuclear war more likely.

There doubtless have been and still are people of goodwill who genuinely believe that deterrence is the best we can do in the nuclear age, that it is a kind of compass, which, if followed, will keep us safe. The following relevant observation is attributed to President Abraham Lincoln in Steven Spielberg's film *Lincoln*: a compass will "point you true north from where you're standing, but it's got no advice about the swamps and deserts and chasms that you'll encounter along the way. If in pursuit of your destination, you plunge ahead, heedless of obstacles, you may achieve nothing more than to sink in a swamp."[xxii]

Even granting that deterrence might (for some people at least) be intended to achieve peace, if it is a compass, then blindly following it toward that destination risks sinking us in a nuclear swamp. Or if deterrence qualifies as a chain, it is assuredly no stronger than its weakest link—of which there are many. We have already considered one category of worrisome weak links: tense international situations, such as the Cuban Missile Crisis and Able Archer 83, each of which greatly increased the chances that the other side's intentions could have been misread and the errors of one disastrously misconstrued by the other.

There are more potential breaking points—preventive war, for example. Take George W. Bush's catastrophic invasion of Iraq in 2003. The Bush Administration had claimed that some threats cannot be deterred and must be eliminated by unilateral action if need be. As it turned out, there was no threat to the United States from Iraq, but the

<hr>

xxii I have not been able to confirm whether the real Lincoln actually said or wrote this. (But in any event, it's good advice!)

American invasion of that country opened a huge Pandora's box of Middle East instability and violence.

It may seem like splitting hairs, but it can be useful to distinguish launching a pre-
ventative war, as the United States did against Iraq, and initiating a preemptive one.
The latter is somewhat more acceptable than the former, because a preemptive attack is
defined as occurring when the threat is immediate and grave, whereas starting a pre-
ventative war is more cold-blooded, supposedly designed to prevent war . . . by starting
one. Tellingly, in some instances, it is unclear which is actually the case. Historians will
long argue whether Israel was acting preemptively or preventatively when it attacked
its Arab neighbors in 1967, allegedly fearing a near-term attack but one that was not
immediately anticipated.

Criteria for preemptive attacks are often thought to involve "anticipatory self-
defense," as historically set out nearly two centuries ago by US Secretary of State
Daniel Webster. An American steamship, the *Caroline*, had been attacked by British
forces in 1837 and then sent ablaze over Niagara Falls to preclude Americans from
using it to support rebels then fighting in Canada. The crisis was peacefully resolved by
diplomacy, whereupon Webster wrote that in order for an act to count as self-defense
(which the British claimed as justifying their actions), it must be "instant, overwhelming, leaving no choice of means, and no moment for deliberation." Moreover, it must be
appropriate to the circumstance, including "nothing unreasonable or excessive, since
the act, justified by the necessity of self-defense, must be limited by that necessity and
kept clearly within it."[73]

According to Professor Leonard Freedman in his book *Deterrence*, the United States
no longer sees itself bound by Webster's analysis, and instead relies on one promulgated by Secretary of War Elihu Root in 1914: "the right of every sovereign state to
protect itself by preventing a condition of affairs in which it will be too late to protect
itself." That understanding would seem to establish a loophole large enough through
which a trigger-happy country could fire a salvo of first-strike missiles.

Wars usually start for many interlocking reasons, not one of which is necessarily
paramount, but that combine to produce the outbreak of hostilities as a final common
path. The First World War is an iconic example that was initiated by an array of causes,
of which the assassination of the Austrian Archduke in 1914 was merely the initiating
match. As tensions increased and all sides worried they might be slower than their
opponents in mobilizing their armies, preplanned orders and schedules fed upon each
other, quickly generating an irresistible drive toward the "war that nobody wanted."
Like a series of one-way ratchets, the process could only move toward hostilities.

In our current world of extended nuclear deterrence, the intensification of a conventional war poses similar dangers. Consider that a US ally, sheltering under our
"nuclear umbrella"—were to be emboldened to think it can escalate a conflict in a way

that otherwise wouldn't happen, because the threat of American nuclear retaliation would keep its opponent from responding too vigorously. An upsetting possibility, for example, would be Saudi Arabia attacking Iran, assuming that the United States would back them up, at least insofar as inhibiting a full-scale Iranian response.

There is a checkered history of America's allies nervously eyeing the prospect that the US nuclear umbrella might not have been available to deter an attack by the Soviet Union in the past, or by Russia in the present. On the other hand, within those countries allegedly enjoying the benefits of extended deterrence from the United States, many concerned citizens—and more than a few national leaders—worry that the extended nuclear umbrella could morph into a lightning rod instead. US nuclear facilities in Germany, for example, have long been targeted by Soviet/Russian planners.[74] In summer 2019, it was "revealed" that US nuclear weapons were stored in Belgium, Germany, Holland, Italy, and Turkey. This had long been known to aficionados of NATO, and certainly by Kremlin military planners. What was notable was not the revelation, but the uproar that it generated among the European public, many of whom had not been aware of their supposed umbrella and were understandably unenthusiastic about being on the receiving end of an atomic rainstorm.

Perhaps certain war-promoting leaders within the United States will someday conclude that, because we have this allegedly potent deterrent, we can act militarily without fear of an effective response. In the event of an ongoing conventional war (regardless of whether it is declared officially), there is a risk that military leaders of any nuclear-armed country on the losing end of a conventional war will be hard pressed to accept defeat when they still possess powerful and allegedly decisive weapons, still unused. They might alternatively attempt to rescue their situation by increasing the combat intensity, hoping either to win by that technique alone or by inducing the other side to back off, fearing yet further escalation to strategic nuclear destruction.

Paradoxically, even the winning side might feel pressed to be the first to cross the nuclear threshold, if it concludes that the loser, having developed weapons and tactics adjusted to the waging of "limited" nuclear war, is about to cross the "firebreak" separating conventional from nuclear. A prolonged stalemate could exert similar pressures. Accordingly, escalation would not be unlikely whenever conflicts break out—perhaps especially among countries supposedly relying on nuclear deterrence, and regardless of who seems to have the upper hand on the battlefield.

Nuclear confrontations could also begin indirectly, either via a proxy war involving allies of a nuclear-armed state or from unanticipated consequences of issuing threats. In 1954, US Secretary of State John Foster Dulles offered two atomic bombs to France: one for use in the defense of Dien Bien Phu, where a French garrison was besieged by the revolutionary nationalist Viet Minh, and the other for use against

China, near the Vietnamese border. In 1958, the National Security Council tentatively approved the use of nuclear weapons to help Chiang Kai-shek (whose government had retreated to Taiwan after being defeated by Mao's communist forces) hold the islands of Quemoy and Matsu, threatened by the mainland Chinese government of Mao Zedong. During the Berlin crisis of 1961, NATO supreme commander Lauris Norstad favored the use of nuclear weapons against the USSR.

When the Marine base at Khe Sanh was surrounded and under siege in 1968 during the Vietnam War, President Johnson asked General Earle Wheeler, Chairman of the Joint Chiefs of Staff, about a possible nuclear defense, and General William Westmoreland, commander of US military forces in Vietnam, established a working group to evaluate this prospect. He became convinced that "the use of a few small tactical nuclear weapons in Vietnam—or even the threat of them—might have brought the war to an end."[75] It seems likely that the use of nuclear weapons in Vietnam was forestalled largely by a different kind of threat: that of adverse public opinion.

Even if the major nuclear powers remain wary of each other, in part because of the risk of escalation, the world is currently rife with other conflict zones that could erupt and either go nuclear themselves or embroil other countries in a "catalytic war," especially if one or both participants are proxies for nuclear-armed ones. Compared to those potentially involving the United States, Russia, and China, many such disputes are more immediate, clear-cut, and impassioned, involving countries that are directly adjacent and disputing hot-button patches of real estate: North and South Korea, India and Pakistan, China and Taiwan, Israel and its Arab neighbors. Moreover, Israeli conventional forces, for example, can only be pushed back a few miles before Jerusalem, Haifa, or Tel Aviv is directly threatened, which increases the likelihood that nuclear authority would be delegated to forward-based field commanders. With their backs to the wall, countries might readily jump across the nuclear firebreak or demand that their allies do so on their behalf.

For example, on October 7, 1973, the second day of the 1973 Yom Kippur War, when Arab states had surprised Israel with an initially successful coordinated attack, Israel is widely acknowledged to have equipped its French-made MD-660 missiles—capable of reaching Cairo or Damascus—with nuclear warheads in case her conventional forces were unable to stem the Egyptian advance. On October 13, US intelligence reported that the Soviet Union responded by shipping nuclear warheads from its naval base at Odessa to Alexandria. It isn't entirely clear that this was accurate: subsequent analysis suggests that it may have resulted from a faulty American radiation detector. If so, this event points to an additional argument against "dual-capable" missiles, in that it is possible that conventional armaments might be mistaken for nuclear, thereby conceivably precipitating nuclear war when one side believes that nuclear use is imminent when it isn't.

Russian Scud missiles, able to carry these warheads to Israeli territory—although incapable of being precisely aimed—were already in Egypt. The United States, in turn, issued a worldwide red alert; it is possible that the Watergate crisis, then gaining steam, was an added factor as President Nixon sought to distract attention from his political difficulties. In any event, at about the same time, with Israel on the ropes and down to only a few days' worth of tank ammunition, Nixon began a massive airlift of needed military supplies. This aid was crucial to Israel and eventually turned the tide in its favor. There was also little doubt that if pushed far enough, the Israeli government would have used its nuclear arsenal, and it seems likely that the Israeli nuclear alert had been intended at least as much to blackmail the United States into providing assistance as in preparation for actual use. This appears to have been the only historical case of successful nuclear coercion, directed against the United States, and—ironically—by a close ally.

A new danger soon surfaced. Israeli forces encircled the Egyptian Third Army and could have destroyed it, leaving Egypt defenseless, whereupon the Soviets placed seven divisions on alert, threatening to intervene on Egypt's behalf. The United States, its earlier red alert having been canceled, responded to this provocation with yet another alert (of the sort currently known as DEFCON 2). This time, however, decentralizing procedures needed for actual combat were not initiated. On the same day, the United Nations passed a resolution calling for a cease-fire and a peacekeeping force, after which the US once more canceled its nuclear alert, but not before the catalytic potential of local wars had been made terrifyingly clear, despite the claims of deterrence advocates.

The most unlikely scenario for deterrence failing is the one most commonly imagined by people unfamiliar with the actual dangers—namely, a bolt-out-of-the-blue, or "boob," attack, in which some evil leader suddenly strokes his handlebar mustache and intones, "Today I shall destroy my enemy, and maybe the world." It is at least plausible that a significant factor preventing such an event is, in fact, deterrence—a case in which the least likely eventuality may be inhibited by an otherwise unjustified contingency. Far more likely antecedents to war involve deterrence failing under a number of scenarios, many of them resulting from problems inherent in deterrence itself.

In addition to first-strike pressure under crisis conditions (a compulsion made more demanding by the advent of counterforce), there is the risk of nuclear Chicken, which we already examined. In such a confrontation, with each country's "driver" confident of the deterrent strength of its vehicle and determined that the other must and will swerve, a devastating collision is possible between the United States and Russia or China, North Korea and the US, China and Russia, and Israel and whichever of its Arab antagonists develops a nuclear arsenal.

Nuclear false alarms are also especially worrisome because, combined with anxiety about being victimized by a first strike, and with the addition of increasing feasibility, at least in theory of such an attack, the report of a missile launch is liable to be taken seriously. We already described the Black Brant Scare, which, although occurring at a time of exceptionally low US–Russia tensions, was nonetheless accorded plausibility by Kremlin authorities. Add to this shortened response times, typically measured in minutes, and the chance of a real response to a false report becomes unacceptably high, especially if an atmosphere of deep distrust is added to the mix.

Importantly, unlike email programs that typically include the option of canceling a message sent accidentally, nuclear "delivery systems" do not have a recall code. Once the SEND command to launch is given, neither ICBMs nor submarine-launched ballistic missiles have any self-detonating system that would, on command, redirect or blow them up in the air. After nuclear bombers are directed to attack their preprogrammed target, there is similarly no code to recall them. Such irrevocability is mandated by the fact that if an "abort" option existed, it could in theory be hacked by an opponent, thereby undermining deterrence. The result is yet another situation in which defending deterrence in the interest of national security ends up undermining actual safety.

Literally hundreds of false alarms and concomitant close calls arose during the Cold War alone, most of which were not made public. Here are a few that were. On October 5, 1960, US early warning radar in Thule, Greenland, reported dozens of Soviet inbound missiles. The North American Aerospace Defense Command (NORAD) went to maximum alert. By good fortune, Soviet Premier Khrushchev was in New York at the time to address the United Nations, so an attack seemed unlikely. It was later revealed that the radar had been fooled by a moonrise over Norway.

On November 24, 1961, Strategic Air Command headquarters lost contact simultaneously with both NORAD and with the Ballistic Missile Early Warning System, interpreted as either a highly unlikely coincidence or a Soviet attack. As it happens, a malfunction of a relay station in Colorado caused that particular multisystem communication breakdown.

On November 9, 1965, a massive power surge and blackout in the northeastern states caused nuclear bomb detectors near several major US cities to light up, leading the Command Center of the Office of Emergency Planning to go on full alert.

On May 23, 1967, unusual solar activity—ejection of coronal material plus huge solar flares—interfered with many radar stations throughout North America, simulating a high-altitude blast, producing an electromagnetic pulse that might have been intended to jam those radars and thus been a prelude to a nuclear attack. Nuclear-armed B-52s were on the verge of taking off to retaliate.

On November 9, 1979, NORAD computers showed that missiles launched from a Soviet submarine were attacking the United States. Fighter–interceptors were scrambled and missile bases went on heightened alert. When no actual attack occurred after six minutes, it was concluded the report was bogus. An investigation found that a technician had accidentally loaded a war games tape into a computer, which, according to *The New York Times*,

> simulated a missile attack on North America, and by mechanical error, that information was transmitted into the highly sensitive early warning system, which read it as a "live launch:' and thus initiated a sequence of events to determine whether the United States was actually under attack.[76]

It wasn't.

On June 3, 1980, President Jimmy Carter's National Security Adviser, Zbigniew Brzezinski, was awakened in the early hours of the morning with the news that 2,200 missiles were within minutes of hitting the US mainland. As he debated whether to phone the president, Brzezinski got another call, that the missiles didn't exist. Brzezinski later reported that he hadn't awakened his wife, not wanting to terrify her unnecessarily when she wouldn't have more than a few minutes to live and nothing to do about it anyhow. According to the National Security Archive, it turned out that a defective computer chip, costing forty-six cents, was to blame.[77] Within months, there were several more false alarms resulting from faulty chips. Faced with this astonishing situation, General David C. Jones, Chairman of the Joint Chiefs of Staff, was not at all abashed. Instead, he took it as an occasion to lecture the Soviets. Their leaders "had better know that we are ready," warned the general, "and that we can respond in a very few minutes."[78] He might have added that we are so ready that we might even respond with no provocation.

The Soviets—now, Russians—have been close-mouthed about their false alarms, although we already noted the erroneous report in 1983, which Stanislaw Petrov diagnosed correctly. It occurred because a newly installed Soviet radar misread an unusual meteorological event, when the sun's rays were reflected by the top of a unique cloud layer. Because of their penchant for secrecy, we know very little about the Soviet/ Russian history of false alarms. But because their radar and computer systems are generally less advanced than those of the United States, it is likely that their error rate is, if anything, higher than ours.

It has been estimated that, since the dawn of nuclear deterrence, there have been not hundreds, but more than a *thousand* false alarms, although most have evidently been caught early so as not to be quite so unsettling. It is probably unrealistic to think that the record of these and other nuclear close calls, in themselves, will have much impact on those people unrelentingly committed to deterrence. Consider that when Hitler's forces precipitated the Battle of the Bulge in 1945, taking the Allies by complete surprise, this was the fourth time in recent history that German armies had invaded France through the Ardennes (the other three being in 1870 during the Franco-Prussian War, in 1914 during World War I, and in 1940, early in World War II). One thing we learn from the study of history is that people don't learn from the study of history.

Other blunders, dangerous in themselves—albeit less liable to precipitate war— have involved accidents with nuclear weapons, known in US military parlance as "broken arrows." These, too, have been distressingly frequent, involving not only cargo planes transporting nuclear weapons, but also the historical gamut of US aircraft, including even fighter–interceptors such as Air Force F-86 Sabre jets and Navy A-4E Skyhawks.

Most, not surprisingly, involved strategic bombers. At different times, B-29s, 36s, 47s, 50s, 52s, and 58s carrying nuclear bombs have crashed, collided with refueling tankers, caught fire on the ground, and literally dropped bombs—by accident—onto US territory. No similar accidents have occurred (or been reported) for B-1s or B-2s. Some examples of broken arrows include a B-36 dropping a nuclear weapon near Kirtland Air Force Base, New Mexico, on May 22, 1957; a B-47 crash near Homestead Air Force Base, Florida, on October 11, 1957; a bomb accidentally released near Mars Bluff, South Carolina, on March 11, 1958; a B-47 crash near Dyess Air Force Base, Texas, on November 4, 1958; and another on November 27 of that year near Chennault Air Force Base, Louisiana. And that's just some of the notable ones during 1957 and 1958 alone. Comparable events continued through the 1960s, '70s, '80s, and '90s, with the frequency declining as airborne patrols involving fully armed nuclear weapons were gradually phased out.

There have been comparatively few incidents involving land-based ICBMs and submarines, the most notable being an event at the Damascus Air Force Base outside of Little Rock, Arkansas, when, on September 18, 1980, a maintenance technician dropped a socket wrench into a Titan missile silo. It pierced the fuel tank, eventually causing a conventional explosion that tossed the missile's W-53 warhead (at nine megatons—about *eight hundred times* more powerful than the Hiroshima bomb) one hundred feet outside the base perimeter. The warhead itself didn't go off. The whole harrowing episode, along with an account of other, similar events has been described in gripping book-length detail.[79]

In many of these cases the high-test explosives surrounding the bomb or warhead cores have detonated, resulting in spectacular explosions and spreading limited radioactive contamination, but so far, the built-in fail-safe mechanisms on the "pits" themselves have worked, and there haven't been any accidental nuclear explosions. Were this to happen, the impact on the immediate victims would be indistinguishable from war. The likelihood, however, is that war itself wouldn't result, although this assumes that the source of the disaster could be determined quickly and the information sent upstream to the National Command Authority so that any tendency to respond precipitously—a serious possibility given the inevitable panic and uncertainty—could be resisted.

Just as we know very little about the history of Soviet/Russian false alarms, we are also largely in the dark concerning their broken arrows, although given that strategic bombers are only a small part of their triad (ICBMs, submarine-launched missiles, and bombers), whereas aircraft comprise the bulk of US nuclear weapons accidents, the Soviet/Russian accident rate has probably been less than ours. There have, however, been a number of acknowledged Soviet/Russian submarine mishaps. Whether these vessels were nuclear armed has been difficult to ascertain.

Next, the question of unauthorized use. Nuclear war would be equally bad whether started by "legitimate" authority or by one or more underlings not authorized to do so. However, it seems reasonable that the more centralized the war-starting authority, the less likely an accidental war, if only because there would be fewer possibly weak links in the command chain. So it would seem that the fewer the number of individuals involved, the better. But here is the catch: insofar as retaliatory capability resides in the president and only the president, a first strike that obliterates them and the briefcase containing the retaliatory codes (i.e., a surprise "decapitating" attack) would also obliterate the prospect of retaliation. Hence, it obliterates deterrence too.

The survival of democracy itself is among those things threatened by deterrence. This goes beyond the obvious concern that, in a world of short flight times and hair-trigger missiles demanding near-instantaneous responses, there is no opportunity for Congress to weigh in on the most consequential of all human decisions, not to mention the constitutional mandate that war requires a vote by Congress. Call it "the delegation dilemma." Official policy, at least in the United States, is that nuclear attacks can only be ordered by the highest civilian government authority—namely, the president. Hence the widely promulgated belief that the coded authorization to launch nuclear war resides solely in a briefcase—the so-called "football"—carried by a trusted aide, always by the President's side.

Bear in mind, however, that for deterrence to work, a would-be attacker must think that a possible first-strike attack would lead to unacceptable (second-strike) retaliation—a conviction that is supposed to inhibit such an attack in the first place. But if such a first strike knocks out the President, the aide, and the football, then retaliation would be precluded along with deterrence.

According to Daniel Ellsberg, who worked for years in the Defense Department as well as the RAND Corporation and was intimately involved in nuclear war-planning scenarios,[xxiii] to solve this problem, the United States has, for decades, delegated nuclear launch authority to an array of military officers, sometimes going down the

[xxiii] Mr. Ellsberg also became famous (for some people, infamous) for releasing, in 1971, the Pentagon Papers, which detailed hitherto secret military documents that described the many lies underpinning US involvement in the Vietnam War.

chain to individual base commanders, occasionally no higher than a major.[80] Since the 1950s, however, the fact of delegation has been officially denied, because it goes against the near-universal expectation that lower level officers should never be capable, on their own, of destroying the world. Hence, the US government adamantly refuses to acknowledge the reality of delegation, which, if known, would be more than a little upsetting to many people, because it contradicts the basic tenet that in, our democracy, the military is subordinate to civilian authority. Official policy, therefore, embraces the fiction that launch authority resides only with the President. Although, if true, this would profoundly destabilize our old friend deterrence, because it could lead to the expectation that a decapitating first strike—one that eliminates the possibility of retaliation—might be feasible.

Delegation: damned if we do; damned if we don't.

The government's refusal to acknowledge delegation once more brings to mind the movie *Dr. Strangelove*, in which the Soviets sought to protect themselves by developing and deploying a doomsday device that would automatically incinerate the planet if a nuclear weapon were detonated on their country. But they didn't tell the United States having planned to announce this later. In reality, nuclear war-fighting authority has probably been delegated in Russia too— so as to preserve deterrence—but this is nonetheless denied in order to preserve something else: their illusion of civilian-dominated control. *Dr. Strangelove* also offers a dark delegation twist involving the United States. After a rogue US Air Force general orders a nuclear attack on the USSR, President Muffley complains that he was supposed to be the only one authorized to order the use of nuclear weapons. He is told that, although he has the sole "authority" for their use, others down the chain of command have the "ability" to do so. Why? "To discourage the Russkies from any hope that they could knock [you] out . . . and escape retaliation."

To understand the delegation dilemma further, it helps to take a quick look at the strategic triad. ICBMs are essentially immobile,[xxiv] which is why their accident rate is quite low. They also have the most reliable communications; hence, they are the most reliably protected against unauthorized firings. However, even ICBMs have a crucial human component: silo-based missile launch officers who are kept from firing their weapons by "permissive action links." These aptly named PALs do us all a great favor by preventing unauthorized use until the correct numerical code is inserted into the electronic system. This code can only be obtained after a PAL unlock has been

---

[xxiv] At least in the United States, although for a time in the 1980s, it was debated whether a next-generation missile, the MX, should be rendered mobile, on trains. North Korea has been reported to deploy mobile ICBMs, as does Russia, on occasion.

authorized from higher up the command chain, after which the missiles and warheads can be armed.

For B-52s, the "go code" provides similar capability. Until this is received, nuclear weapons cannot be armed and, when it has arrived, the pilot, radar navigator, and electronics warfare officer must all cooperate. For submarines, the situation is less clear. To be invulnerable against attack (their chief selling point in the deterrence scheme), nuclear-armed submarines must remain submerged; but, while deep underwater, they have great difficulty receiving or sending messages. Greater depth is especially advisable during heightened tensions, when antisubmarine warfare would be particularly likely; and yet, greater depth means greater difficulty in receiving messages. Accordingly, it is almost certain—although never officially confirmed—that unlike ICBMs and strategic bombers, submarines can fire their nuclear weapons without transmission from the outside. That is, each sub likely has the capacity to initiate nuclear war by itself.

Inside each submarine, several "reliable"[xxv] crew members, including the captain, first officer, and electronics warfare officer, must act together to arm and fire their missiles. Errors seem unlikely, but are not impossible. On one occasion, a communications specialist misread a text calling for a surprise drill as the real thing: the dreaded Emergency Action Message. Fortunately, the captain caught the error, but then he did something interesting. To test the reliability of his officers, he kept the error secret, allowing everyone but himself to proceed as though it actually *was* the real thing. His crew did as they had been trained and ordered, whereupon the captain, satisfied, ended the suspense by revealing the truth. This little drama ended well.

But what if the captain had died of a heart attack or stroke during the "drill," leaving his crew committed to starting a nuclear war? This is not so far-fetched. Pilots, bus drivers, and train conductors have died at their controls, and the results, although tragic, have been mild by contrast. Note also that there were nine *reported* collisions between US nuclear submarines and "hostile vessels" in Soviet waters during the 1970s alone, five of which involved Soviet subs thought to be nuclear armed.

Drug use is officially prohibited in every nuclear weapons context. But, in 1969, a Cuban defector flew his MIG safely to mainland Florida, right through the coastal radar defenses, without arousing any alarm. The resulting investigation of Nike-Hercules missile crews resulted in the arrest of thirty-five men who were assigned to these crews. They were charged with using drugs, including LSD, which can cause not only inattentiveness, but also delusions and hallucinations.

[xxv] It is acknowledged by people with experience in the US military's Human Reliability Programs that Donald Trump would almost certainly be rejected for possible nuclear duty, and not just because of age.

Given these and a few other circumstances leading to close calls, it is noteworthy that nuclear weapons haven't been used in anger since August 1945. This might result from the wisdom of deterrence. Or something else. Former Defense Secretary Robert McNamara—no peacenik—said the following in his last address describing our deliverance from the Cuban Missile Crisis:

> I was there; I have had direct experience in trying to handle a nuclear crisis with the fate of the world on the line; and because of my experience, I know—I am not guessing or speculating. I know that we were just plain lucky in October 1962—and that without that luck most of you would never have been born because the world would have been destroyed instantly or made unlivable in October 1962. And something like it could happen today, tonight, next year. It will happen at some point. That is why we must abolish nuclear weapons as soon as possible.[81]

As soon as possible just might be soon enough. Or maybe too late. When Adelaide, Nathan Detroit's long-suffering fiancée in the Broadway musical *Guys and Dolls*, is told by Nathan that they'd get married "sooner or later," she replies, "It is already too late to be sooner. And if it gets much later, soon it will be too late even to be later."

A PRESCRIPTION

A primary goal of this book has been not only to describe the use of threats in animals, and their use and abuse by people, but also to critique the widespread human use of deterrence. Paradoxically, therefore, I hope to deter deterrence! If I have succeeded, then by this point you might realize that nuclear deterrence is particularly undeserving of the hushed, unexamined respect that it typically enjoys, and that it is more primitive, less proportionate, less ethical, less reliant on a consistently rational foundation—and immensely more dangerous than most people realize.

Beyond its many downsides, the most pernicious skeleton rattling in the closet of nuclear deterrence—and the most cogent reason to be skeptical of the whole enterprise (short of its potential for catastrophic failure)—is that it has been and continues to be the positive face, the bedrock legitimation for the ongoing development, deployment, maintenance, and escalation of the whole nuclear weapons enterprise. It is because of misrepresentations tied to nuclear deterrence that we live under what President Kennedy called a "nuclear sword of Damocles," held by a slender thread, liable to break at any time. Urgently needed, therefore, is recognition that when it comes to national security, nuclear weapons are the problem, not the solution. And nuclear deterrence is the phony justification for keeping this problem around.

Moreover, the conventional forces of the United States greatly exceed those of any other country—and by a wide margin. Insofar as nuclear weapons are something like the "great equalizer" that the Colt 45 was on the Western frontier of the US, getting out from under the nuclear threat would, rather than weakening the international clout of the United States, actually increase it. Outgrowing deterrence would be more like abolishing slavery, a parallel that nuclear devotees are likely to dismiss because, although getting rid of slavery affected only the wealth of slaving nations, eliminating nuclear weapons would remove an allegedly irreplaceable fundamental guarantee of the possessor state's security. But every government that abolished slavery (including that which became, briefly, the Confederate States of America) eventually discovered that their way of life really didn't depend on it, and that there were alternatives.

The Stone Age didn't end because we ran out of stones, but because we moved on to other ways of doing things.[xxvi] When it comes to nuclear deterrence, there are other ways of doing things.

A few years ago, a bizarre and profoundly unpleasant thing was discovered lurking in the sewers of London. Dubbed the "fatberg," it was a solid 130-ton glob of hardened glop, composed mostly of congealed cooking fat that had been poured down English drains, which served as a coagulating nucleus that captured an array of equally nasty materials (condoms, baby diapers, toilet paper), preventing the free flow of water and posing a public health menace. It was eventually dismembered, although not without considerable difficulty. Deterrence is the fatberg of national security, accumulating the detritus of career advancement, money-hungry corporations, and ideologues, while blocking the free flow of strategic imagination and innovation. Perhaps it, too, will be dismembered.

Doing so will require reconfiguring our attitude not only toward deterrence, but also toward our competitors (to be distinguished from genuine enemies). Shakespeare's play The Winter's Tale includes this famous stage direction, which generations of theater professionals have struggled to represent: "Exit, pursued by a bear." Rethinking deterrence would require going beyond decades of assumptions—notably, about our being lethally pursued by the Russian bear. Although the Cold War is officially over, it persists, especially via Russo-phobia—plus Sinophobia, Islamophobia, Irano-phobia, North Korea-phobia and their ilk—which influence and deform the thinking of nearly all of today's strategic analysts and weapons procurers.

Toward the end of his life, Paul Nitze, who had served many administrations as one of the most unrelentingly hawkish members of the policymaking nuclear priesthood,

---

[xxvi] I am pretty sure that someone came up with these bons mots before me, but haven't been able to ascertain who, or when,

came to just this recognition. In an op-ed piece titled "A Threat Mostly to Ourselves," he wrote the following:

> I can think of no circumstances under which it would be wise for the United States to use nuclear weapons, even in retaliation for their use against us. What, for example, would our targets be? It is impossible to conceive of a target that could be hit without large-scale destruction of many innocent people. . . . In view of the fact that we can achieve our objectives with conventional weapons, there is no purpose to be gained through the use of our nuclear arsenal. To use it would merely guarantee the annihilation of hundreds of thousands of people, none of whom would have been responsible for the decisions invoked in bringing about the weapons' use, not to mention incalculable damage to our natural environment.[82]

General Douglas MacArthur—like Nitze, no shrinking violet when it came to the question of military threats and how to respond to them—was outraged by policies that exaggerated external threats. He lectured the annual stockholder's meeting of the Sperry Rand Corporation[xxvii]:

> Our government has kept us in a perpetual state of fear—kept us in a continuous stampede of patriotic fervor—with the cry of grave national emergency. Always there has been some terrible evil at home or some monstrous foreign power that was going to gobble us up if we did not blindly rally behind it.[83]

In *The Ambassadors*, Henry James described Paris as "like some huge iridescent object, a jewel brilliant and hard, in which parts were not to be discriminated nor differences comfortably marked. It twinkled and trembled and melted together, and what seemed all surface one moment seemed all depth the next." Nuclear deterrence has been sold to the American people as a strategy of great depth, and yet, it is actually all surface. Recall that well-known wizard who bellowed, "I am Oz, the great and terrible," to which a certain little girl replied, trembling, "I am Dorothy, the small and meek." But later, when her dog Toto tips over a screen, we see that the wizard is just a little fellow with a bald head and wrinkled face, who cries, "Pay no attention to the man behind the curtain!" He, too, was all surface.

Deterrence isn't only all surface; it is also—and this is key—like the fatberg, a troublesome creation of human beings but something therefore subject to human

---

[xxvii] Then, as now, a major military contractor.

intervention. In this sense, it accords with the second stanza of William Blake's rueful poem, "London," in being a product of our own minds:

> In every cry of every Man,
> In every Infants cry of fear,
> In every voice: in every ban,
> The mind-forg'd manacles I hear.

The current book has been mostly descriptive, seeking to examine, illuminate, and then usually to criticize a number of conspicuous threats, ending with the mind-forg'd manacles of deterrence. It will now conclude with some prescription, assuming that although threats will never go away entirely, they can and should be used wisely and definitely not abused willy-nilly. Actionable wisdom in this regard means recognizing that conveying threats is often "natural," and therefore frequently feels good—at least to the threatener—but that natural does not necessarily equate to good. Hurricanes, tornados, earthquakes, tsunamis, cholera, AIDS, tuberculosis, and COVID-19 are all (mostly) natural, but far from good. By the same token, doing what comes naturally is often easy, but not necessarily admirable.

Fortunately, however, life doesn't consist only of threats. There is cooperation, love, creativity, and innumerable technicolor delights, blooming and buzzing confusion, and tragedies too. Although it may always be necessary to maintain order, interpersonally and intrasocietally no less than internationally, it assuredly is not necessary to base so much of our lives on the conveyance of and response to threat—that of nuclear annihilation most especially. This makes it all the more important to move beyond hardening of the categories, to unblock the human imagination, and to minimize the self-inflicted wounds of threats that need not and should not exist. It's a tall order, but as T.S. Eliot once wrote, unless you are in over your head, how do you know how tall you are?

There is a story, said to be of Cherokee origin, in which a young girl was frightened by a recurring dream in which two wolves viciously fought each other. She described this to her grandfather, a tribal elder, renowned for his wisdom, who explained that everyone has both peaceful and violent wolves within them, and they struggle for control. At this, the child was even more worried and asked who wins. Her grandfather replied, "The one you feed."

How might the nuclear, deterrence-based fatberg be dissolved, the mind-forg'd manacles unlocked, the wolf be starved? Risking cliché or copout, this is a subject that warrants its own separate and detailed treatment, if only because imagining one's own death may well be easier than imagining a life without the fatberg, or how to starve the right wolf. But for now, here is an abbreviated diet plan. It does not presume humanity evolving to a higher, more spiritually admirable level. It does not demand that lions lie down with lambs or that universal love replace realpolitik. It does not rely upon rainbows or sparkly unicorns.

First, recognize and promote the fact that nuclear weapons are unacceptably perilous. Second, confront the claim that abandoning nuclear deterrence would itself constitute a dangerous threat, by emphasizing the much greater threat that it currently poses, the fact that it has been credited with success that it hasn't achieved, and pointing out there is no military mission to which nuclear weapons are currently assigned that could not be performed more effectively, if need be, and with less devastating "collateral damage" to the planetary biosphere with conventional weapons—a dimension in which the United States is the world's unquestioned leader. In short, disenthrall ourselves from the claim that nuclear weapons and deterrence are good for us and for the world.

Here are some specific things to do, in the service first of fixing deterrence, then overhauling it, and, finally, getting rid of it altogether:

- Take all existing nuclear weapons off hair-trigger alert and separate nuclear warheads from their delivery systems to reduce the risk that any other country might perceive they constitute a sudden, first-strike threat to them. This would also reduce any temptation to use them.
- Decommission all land-based ICBMs. These sitting ducks could induce another country (fearing a first strike by the United States) to launch its own first strike in the event of an international crisis, and—because of the vulnerability of these missiles, the precise locations of which are precisely known—induce US military and civilian leaders to fear just such a first strike, and possibly preempt it.
- Announce and institutionalize a no-first-use policy. Most Americans do not realize that "first use" has long been and continues to be official US policy. As a result, deterrence, bad as it is, is not necessarily what it seems; to a large extent, reality is even worse. Abandoning any prospect of first use would have calming international implications of the weapons deployed and the circumstances calling for their use.
- Eliminate any hints of launch-on-warning, especially anything that gives computers ultimate control.
- Halt all plans for "modernization" of the nuclear arsenal, excepting those that enhance safeguards and reduce the likelihood of false alarms, accidental launches, and so forth.
- Eliminate all tactical, battlefield nuclear weapons.
- Rejoin the Antiballistic Missile Treaty to eliminate other countries' fear that we might be so foolish as to launch an attack, thinking that we would be immune to retaliation. Although it is theoretically possible that some day a genuine missile defense system could render nuclear weapons "impotent and obsolete," thereby

rendering deterrence irrelevant, such an achievement would have to jump through many independent hoops, each very narrow, and all of which would have to be successfully navigated simultaneously for the scheme to work: (1) defy the problem of physics inherent in successfully intercepting high-speed incoming missiles; (2) solve the challenges of small, radar-evading multiple warheads, chaff, and other anti-ABM measures, which are much more easily deployed than any realistic missile defense; (3) satisfactorily share the technology with other countries, some of them adversaries; and (4) overcome the skepticism of governments, scientists, strategists, and the public during the course of deployment and subsequent maintenance. On balance, it would be much easier and safer to simply get rid of nuclear weapons!

- Rejoin the Intermediate-range Nuclear Forces Treaty. Its abandonment since 2019 risks propelling the world toward numerous regional arms races, notably with China as well as with Russia.

- Rejoin the Joint Comprehensive Plan of Action (commonly known as the "Iran Nuclear Deal"), to which, prior to the Trump Administration's withdrawal, the Iranian government had scrupulously abided, and which severely hampered its nuclear program.

- Negotiate a follow-on to the New-START Treaty, currently scheduled to expire in 2021. If New-START dies, there won't be any verification or onsite inspections. One result will be rampant worst-case planning, with the risk of rapid accelerated nuclear deployments on all sides. Bear in mind that no country needs anything approaching thousands of nuclear weapons: recall that China, with a population roughly four times that of the United States, has about 250 to 300 nuclear warheads, which itself is too many.

- Sign and ratify the UN Treaty on the Prohibition of Nuclear Weapons, which makes it illegal to "develop, test, produce, manufacture, otherwise acquire, possess, or stockpile nuclear weapons or other nuclear explosive devices." This ongoing effort to make nuclear weapons a violation of international law is initially symbolic and obviously will not be implemented immediately, but it goes a long way toward delegitimating nuclear weapons and the threat they embody.

- In the interim—and insofar as it may be politically necessary to reassure those influential self-styled realists whose "realism" is circumscribed by an unrealistic reliance on nuclear threats—draw-down to a bare-bones minimum arsenal, following the lead of the United Kingdom, France, and China, in the neighborhood of at most a few hundred bombs or warheads. At the same time, actively pursue agreements to verifiably eliminate all nuclear weapons. *All* of them!

- Until this is achieved, end the unchecked authority of the US President, no matter who, to order nuclear war.

- Engage in a vigorous policy of active threat reduction and confidence-building measures with the government of North Korea, as well as with Russia, Iran, and all other potential adversaries.
- Don't be bamboozled by the oft-repeated claim by defense intellectuals that "you can't put the nuclear genie back in the bottle." The physicist Edward Teller ("father of the hydrogen bomb") had urged that they be used to melt arctic ice, thereby freeing up the Northwest Passage, and to dig seaports, while other physicists, including Freeman Dyson, spent years on Project Orion, trying to design a rocket that would be powered by a successive series of nuclear explosions. Crappy ideas don't have to be forgotten to be abandoned.

Ditto for lousy technologies. The earliest high-wheel bicycles—popular in the 1870s and 1880s—had huge front wheels and tiny rear ones, and were not only difficult to ride, but dangerous to fall off. Between 1897 and 1927, the Stanley Motor Carriage Company sold more than ten thousand Stanley Steamers, automobiles powered by steam engines. These technologies are now comical curiosities, reserved for museums. Perhaps transportation intellectuals had warned that "you can't put the Stanley Steamer genie back in the bottle."

Bad genies don't necessarily have to be stuffed back into their bottles; they can simply be left to fall of their own weight.

NOTES

1. Quoted in Preston, D. 2002. *Lusitania: An Epic Tragedy.* New York, N.Y.: Walker & Co.
2. Quoted in Whan, V.E. (ed.) 1965. *A Soldier Speaks: Public Papers and Speeches of General of the Army Douglas MacArthur.* New York, N.Y.: Praeger.
3. https://historynewsnetwork.org/article/173415.
4. See, for example, J. Haidt, J. 2012. *The Righteous Mind: Why Good People Are Divided by Politics and Religion.* New York, N.Y.: Pantheon; Yang, J.Z., Chu, H., and Kahlor, L. 2018. Fearful conservatives, angry liberals: Information processing related to the 2016 presidential election and climate change. *Journalism & Mass Communication Quarterly,* 96(3), 742–766.
5. Brodie, Be. 1946. *The Absolute Weapon: Atomic Power and World Order.* New York, N.Y.: Harcourt, Brace.
6. Oppenheimer, J.R. 1953. Atomic weapons and American policy. *Foreign Affairs, 31,* 202–205.
7. Quoted in Green, R. 2018. *Security Without Nuclear Deterrence.* New York, N.Y.: CreateSpace Independent Publishing.
8. Wittgenstein, L. 2009. *Philosophical Investigations,* 4th ed. New York, N.Y.: Wiley-Blackwell.
9. Hirschbein, R. 1989. *Newest Weapons/Oldest Psychology: The Dialectics of American Nuclear Strategy.* New York, N.Y.: Peter Lang.
10. Prins, G. (ed.). 1984. *The Nuclear Crisis Reader.* New York, N.Y.: Vintage.
11. Joseph Cirincione, personal communication.

12. Skinner, B. F. 1948. 'Superstition" in the pigeon. *Journal of Experimental Psychology, 38*(2), 168–172.

13. Hellman, M. 2019. *Rethinking National Security*. Washington, D.C.: Federation of American Scientists Special Report .

14. 1897. Terrible automatic weapons of war. *The New York Times*, March 28.

15. Sechser, T. and Fuhrmann, M. 2017. *Nuclear Weapons and Coercive Diplomacy*. New York, N.Y.: Cambridge University Press.

16. Ibid.

17. Wilson, W. 2013. *Five Myths About Nuclear Weapons*. New York, N.Y.: Houghton Mifflin Harcourt.

18. https://inmenlo.com/2012/08/23/slac-theoretical-physicist-sidney-drell-reflects-on-his-public-policy-work-involving-nuclear-disarmament/.

19. Sagan, S. 1995. *The Limits of Safety*. Princeton, N.J.: Princeton University Press.

20. Boorstin, D.J. 1992. *The Image: A Guide to Pseudo-Events in America*. New York, N.Y.: Vintage.

21. Krepon, M. https://www.armscontrolwonk.com/archive/1204746/justice-and-the-bomb?.

22. https://www.nytimes.com/1989/01/29/world/soviets-say-nuclear-warheads-were-deployed-in-cuba-in-62.html. A

23. Quoted in Mozgovoi, A. 2002. The Cuban samba of the quartet of foxtrots: Soviet submarines in the Caribbean crisis of 1962. In W. Burr and T.S. Blanton (eds.). *Military Parade*. http://nsarchive.gwu.edu/NSAEBB/NSAEBB75/.

24. Ellsberg, D. 2018. *The Doomsday Machine*. New York, N.Y.: Bloomsbury.

25. Memorandum to the Director of Central Intelligence, Deputy Director of Central Intelligence from Herbert F. Meyer, Vice Chairman, National Intelligence Council. Why is the world so dangerous?

26. The 1983 war scare: "the last paroxysm" of the Cold War part II. http//:nsarchive.gwu.edu.

27. 1990. The Soviet "war scare." Feb. 15. https://nsarchive2.gwu.edu/nukevault/ebb533-The-Able-Archer-War-Scare-Declassified-PFIAB-Report-Released/.

28. Pry, P. 1999. *War Scare: Russia and America on the Nuclear Brink*. Westport, Conn.: Greenwood Publishing.

29. Baldwin, J. 1963. *The Artist's Struggle for Integrity*. Oasis Digital Press. [Kindle edition.].

30. McNamara, R.S. 1968. *The Essence of Security: Reflections in Office*. New York, N.Y.: Hodder & Stoughton.

31. Turco, R.P., Toon, O.B., Ackerman, T.P., Pollack, J.B., and Sagan, C. 1983. Nuclear winter: global consequences of multiple nuclear explosions. *Science, 222*(4630), 1283–1292.

32. https://www.latimes.com/archives/la-xpm-1985-03-14-mn-26564-story.html.

33. Badash, L. 2001. Nuclear winter: scientists in the political arena. *Physics in Perspective, 3*(1), 76–105.

34. Robock, A., Oman, L., Stenchikov, G.L., Toon, O.B., Bardeen, C., and Turco, R.P. 2007. Climatic consequences of regional nuclear conflicts. *Atmospheric Chemistry and Physics, 7*(8), 2003–2012; Toon, O.B., Robock, A., and Turco, R.P. 2014. Environmental consequences of nuclear war. *AIP Conference Proceedings, 1596*(1), 65–73; Yu, P., Toon, O.B., Bardeen, C.G., Zhu, Y., Rosenlof, K.H., Portmann, R.W., Thornberry, T.D., Gao, R-S., Davis, S.M., Wolf, E.T., de Gouw, J., Peterson, D.A., Fromm, M.D., and Robock, A. 2019. Black carbon lofts wildfire smoke high into the stratosphere to form a persistent plume. *Science, 365*, 587–590; Jonas Jägermeyr et al. 2020. A regional nuclear conflict would compromise global food security. *Proceedings of the National Academy of Sciences*, doi: 10.1073/pnas.1919049117.

35. Colby, E. 2018. If you want peace, prepare for nuclear war. *Foreign Affairs*, 97, 25–32.

36. Kahn, H. 1965. *On Escalation*. New York, N.Y.: Penguin.

37. https://docs.house.gov/meetings/AS/AS00/20150625/103669/HHRG-114-AS00-Wstate-WinnefeldJrUSNJ-20150625.pdf.

38. Krepon, M., n. 21.

39. Scheling, T.C. 1960. *The Strategy of Conflict*. Cambridge, Mass.: Harvard University Press.

40. Schelling, T.C. 1966. *Arms and Influence*. New Haven, Conn.: Yale University Press.

41. 1956. *The New York Times*, February 26.

42. Kissinger, H. 2013. *A World Restored*. New York, N.Y.: Echo Point.

43. West, R. 1940. *Black Lamb and Grey Falcon: A Journey Through Yugoslavia*. New York, N.Y.: Viking.

44. Barash, D.P. and Webel, C.P. 2018. *Peace and Conflict Studies*, 4th ed. Thousand Oaks, CA: Sage Publications.

45. Scarry, E. 2014. *Thermonuclear Monarchy*. New York, N.Y.: W. W. Norton.

46. House Committee on International Relations. 1976. *First Use of Nuclear Weapons: Preserving Responsible Control*. Washington, D.C.: US Government Printing Office.

47. Helfand, I. 2018. https://www.cnn.com/2017/01/18/opinions/nuclear-codes-trump-disarmament-opinion-helfand/index.html.

48. Schelling, T.C. 1967. *Arms and Influence*. New York, N.Y.: Praeger.

49. Ibid.

50. Quoted in Sherry, M. 1987. *The Rise of American Air Power: The Creation of Armageddon*. New Haven, Conn.: Yale University Press.

51. Quoted in Walzer, M. 2000. *Just and Unjust Wars*. New York, N.Y.: Basic Books.

52. https://www.nytimes.com/1986/04/27/us/ban-on-a-arms-urged-in-study-by-methodists.html.

53. Barash, D.P. and Lipton, J.E. 1985. *The Caveman and the Bomb: Evolution, Human Nature and Nuclear War*. New York, N.Y.: McGraw-Hill.

54. Perry, W.J. 2015. *My Journey at the Nuclear Brink*. Stanford, Calif.: Stanford University Press.

55. Barash, D.P. 1986. *The Arms Race and Nuclear War*. Belmont, Calif.: Wadsworth Publishing.

56. Kennan, G. 1981. https://www.nybooks.com/articles/1981/07/16/a-modest-proposal/.

57. https://www.thenation.com/article/us-nuclear-arsenal-triad/.

58. Kroenig, M. 2018. *The Logic of American Nuclear Strategy: Why Strategic Superiority Matters*. New York, N.Y.: Oxford University Press.

59. Dick, P.K. 2011. *Valis*. New York, N.Y.: Mariner Books.

60. Kahn, H. 1984. *Thinking About the Unthinkable in the 1980s*. New York, N.Y.: Simon and Schuster.

61. Barash, D.P. 2004. *The Survival Game: How Game Theory Explains the Biology of Cooperation and Competition*. New York, N.Y.: Henry Holt.

62. Quoted in Scheer, R. 1982. *With Enough Shovels: Reagan, Bush and Nuclear War*. New York, N.Y.: Random House.

63. Schelling, n. 40.

64. Ravenal, E.C. 1982. Counterforce and alliance: the ultimate connection. *International Security*, 6, 26–43.

65. Gerlach, P. 2017. The games economists play: why economics students behave more self-ishly than other students. *PLoS One*, 12(9), e0183814..

66. For an account of these episodes see Herken, G. 1987. The earthly origin of Star Wars. *The Bulletin of the Atomic Scientists*, October, 114.

67. Krepon, M. 2019. https://www.nytimes.com/2019/03/03/opinion/nuclear-weapons-congress.html.
68. http://www.newamericancentury.org/RebuildingAmericasDefenses.pdf.
69. Kohn, R.H. and Harahan, J.P., eds. 1988. *Strategic Air Warfare: An Interview with Generals Curtis E. LeMay, Leon W. Johnson, David A. Burchinal, and Jack J. Catton.* Washington, D.C.: Office of Air Force History United States Air Force.
70. https://theintercept.com/2017/05/03/why-do-north-koreans-hate-us-one-reason-they-remember-the-korean-war/.
71. Quoted in Sang-Hun, C. 2018. North Korea weaponizes its deal with Trump to tangle talks. *The New York Times*, October 12.
72. Quoted in https://www.foreignaffairs.com/articles/north-korea/2017-05-23/atoms-pyongyang.
73. Cited in Arend, A.C. 2003. International law and the use of military force. *The Washington Quarterly*, 26(2), 89–103.
74. Green, R. 2019. https://www.opendemocracy.net/en/new-nuclear-deterrence-and-disarmament-crisis/.
75. Tannenwald, N. 2006. Nuclear weapons and the Vietnam War. *Journal of Strategic Studies*, 29(4), 675–722.
76. Steven, M., and Mele, Christopher Mele. 2018. Causes of false missile alerts: The sun, the moon and a 46-cent chip. *The New York Times*.
77. https://nsarchive2.gwu.edu/nukevault/ebb371/.
78. Quoted in Barash, D.P. and Lipton, J.E. 1983. *Stop Nuclear War! A Handbook.* New York, N.Y.: Grove Press.
79. Schlosser, E. 2013. *Command and Control: Nuclear Weapons, the Damascus Incident and the Illusion of Security.* New York, N.Y.: Penguin Press.
80. Ellsberg, n. 24.
81. Quoted in Blight, J. and Lang, J. 2017. *Dark Beyond Darkness: The Cuban Missile Crisis as History, Warning and Catalyst.* Lanham, Md.: Rowman & Littlefield.
82. Nitze, P. 1999. https://www.nytimes.com/1999/10/28/opinion/a-threat-mostly-to-ourselves.html.
83. Imparato, E.T. 2000. *General MacArthur Speeches and Reports 1908–1964.* New York, N.Y.: Turner.

Appendix A

Death by Deterrence

The following is a selection from "Death by Deterrence," a *cri de coeur* by General George Lee Butler, a retired four-star Air Force General and former Commander-in-Chief of US Strategic Command, where he had responsibility for all US strategic nuclear forces. (This material was first published in Resurgence & Ecologist, issue 193, March/April 1999, and is reprinted here with permission. All rights are reserved to The Resurgence Trust.) In it, General Butler described "a world beset with tidal forces, towering egos, maddening contradictions, alien constructs and insane risks," observing that "the threat to use nuclear weapons is indefensible." He goes on to ask

By what authority do succeeding generations of leaders in the nuclear weapons states usurp the power to dictate the odds of continued life on our planet? Most urgently, why does such breathtaking audacity persist at a moment when we should stand trembling in the face of our folly and united in our commitment to abolish its most deadly manifestation? . . . There is no other way to understand the willingness to condone nuclear weapons except to believe they are the natural accomplice of visceral enmity. They thrive in the emotional climate born of utter alienation and isolation. The unbounded wantonness of their effects is a perfect companion to the urge to destroy completely. They play on our deepest fears and pander to our darkest instincts. They corrode our sense of humanity, numb our capacity for moral outrage, and make thinkable the unimaginable . . . .

Like millions of others, I was caught up in the holy war, inured to its costs and consequences, trusting in the wisdom of succeeding generations of military and civilian leaders. The first requirement of unconditional belief in the efficacy of nuclear weapons was early and perfectly met for us: our homeland was the target of a consuming evil, poised to strike without warning and without mercy. . . . Bound up in this singular term, this familiar touchstone of security dating back to antiquity, was the intellectually comforting and deceptively simple justification for taking the most extreme risks and the expenditure of trillions of dollars. It was our shield and by extension our sword. The nuclear priesthood extolled its virtues and bowed to its demands. Allies yielded grudgingly to its dictates even while decrying its risks and costs. We brandished it at our enemies and presumed they embraced its suicidal corollary of mutually assured destruction. We ignored, discounted or dismissed its flaws and cling still to the belief that deterrence is valid . . . .

Deterrence carried the seed, born of an irresolvable internal contradiction, that spurred an insatiable arms race. Nuclear deterrence hinges on the credibility to mount a devastating retaliation under the most extreme conditions of war initiation. Perversely, the redundant and survivable force required to meet this exacting test is readily perceived by a darkly suspicious adversary as capable, even designed, to execute a disarming first strike. Such advantage can never be conceded between nuclear rivals. It must be answered, reduced, nullified. Fears are fanned, the rivalry intensified. New technology is inspired, new systems roll from production lines. The correlation of force begins to shift, and the bar of deterrence ratchets higher, igniting yet another cycle of trepidation, worst-case assumptions and ever-mounting levels of destructive capability . . . .

I participated in the elaboration of basing schemes that bordered on the comical and force levels that in retrospect defied reason. I was responsible for war plans with over 12,000 targets, many struck with repeated nuclear blows, some to the point of complete absurdity. I was a veteran participant in an arena where the most destructive power ever unleashed became the prize in a no-holds-barred competition among organizations whose principal interest was to enhance rather than constrain its application. And through every corridor, in every impassioned plea, in every fevered debate rang the rallying cry, deterrence, deterrence, deterrence. . . . The exorbitant price of nuclear war quickly exceeded the rapidly depreciating value of a tenuous mutual wariness. Invoking deterrence became a cheap rhetorical parlour trick, a verbal sleight of hand. Proponents persist in dressing it up to court changing times and temperaments, hemming and re-hemming to fit shrinking or distorted threats.

Deterrence is a slippery conceptual slope. It is not stable, nor is it static; its wiles cannot be contained. It is both master and slave. It seduces the scientist yet bends to his creation. It serves the ends of evil as well as those of noble intent. It holds guilty the innocent as well as the culpable. It gives easy semantic cover to nuclear weapons, masking the horrors of employment with siren veils of infallibility. At best it is a gamble no mortal should pretend to make. At worst it invokes death on a scale rivalling the power of the creator.

From the very beginnings of the nuclear era, the objective scrutiny and searching debate essential to adequate comprehension and responsible oversight of its vast enterprises were foreshortened or foregone. The cold light of dispassionate scrutiny was shuttered in the name of security, doubts dismissed in the name of an acute and unrelenting threat, objections overruled by the incantations of the nuclear priesthood.

Sad to say, the cold war lives on in the minds of those who cannot let go the fears, the beliefs, and the enmities born of the nuclear age. They cling to deterrence, clutch its tattered promise to their breast, shake it wistfully at bygone adversaries and balefully at new or imagined ones. They are gripped still by its awful willingness not simply to tempt the apocalypse but to prepare its way. What better illustration of misplaced faith in nuclear deterrence than the persistent belief that retaliation with nuclear weapons is a legitimate and appropriate response to post-cold-war threats posed by weapons of mass destruction. What could possibly justify our resort to the very means we properly abhor and condemn?

As a nation we have no greater responsibility than to bring the nuclear era to a close. Our present policies, plans and postures governing nuclear weapons make us prisoner still to an age of intolerable danger. We cannot at once keep sacred the miracle of existence and hold sacrosanct the capacity to destroy it. We cannot hold hostage to sovereign gridlock the keys to final deliverance from the nuclear nightmare. We cannot withhold the resources essential to break its grip, to reduce its dangers. We cannot sit in silent acquiescence to the faded homilies of the nuclear priesthood. It is time to reassert the primacy of individual conscience, the voice of reason and the rightful interests of humanity.

Appendix B

Deterrence Down on the Farm
(A personal note from the author)

I am not opposed to all uses of deterrence. My wife and I live on a ten-acre horse farm about fifteen miles east of Seattle. It is patrolled by a territorial 140-pound Anatolian shepherd, and we have never been burglarized. We also maintain an electric fence around the perimeter of our property to keep our animals in (including the Anatolian), and others out. Even if we could, however, we wouldn't seek to deter intruders by threatening to blow up the neighborhood. Aside from the insuperable practical and moral constraints, such a threat would lack credibility—as we have seen, this is an enormous, consequential, and unsolved problem when it comes to nuclear deterrence—and also because, on occasion, accidents happen.

We have sometimes unintentionally touched our "hot wire." Although painful, such events simply reinforce our subsequent caution; the effect is unpleasant but far from lethal. Our big dog, more than once, has misfired and, in a fit of excessive, redirected exuberance—for example, when a coyote was tantalizingly close but on the other side of the fence—he has attacked one of our smaller dogs. The outcome has been financially beneficial for our local veterinarian, and the canine victim has always recovered—unlike the all-but-certain consequence of an accidental missile launch or an ostensibly "limited" nuclear war.

My opposition to deterrence is thus less than absolute, and yet herein lies part of the problem when it comes to nuclear deterrence. Although carrots are generally better than sticks, and threats are less effective (also less ethical) than rewards, the fact that threats sometimes work—that, indeed, they are baked into much of the animal and human world—readily leads to the unspoken assumption that what's good, at least occasionally, in the interpersonal and conventional domain is also good when it comes to nuclear weapons.

Nuclear deterrence isn't normally discussed in polite civilian conversation, but start looking for nonnuclear deterrence and you will find it almost everywhere: "Don't make me say this a second time . . . or else." Or "If you hit your sister again, you'll be in a time-out." Strong doors and locks are intended to deter crime. So are police. But people who invest in a home security system are unlikely to install one that responds to a burglary by blowing up the house—even if such a threat is advertised on a conspicuous lawn sign. By the same token, although a policeman on the corner may well deter crime, recent well-publicized events make it clear that sometimes the police use lethal force when it is not called for.

A saving grace is that such occasions, although tragic, do not result in destroying an entire city, country, or planet. Not so in the nuclear world. It is also true, as developed throughout this book, that threats and punishments are widespread in nature as well as among human beings. It should also be clear that the threat of ultimate punishment—in other words, deterrence—has no place in a world of nuclear weapons. Or, more to the point, nuclear weapons have no place in a world suffused with threats and punishment.

To protect ourselves as well as our horses, we have no choice but to train them carefully; a half-ton animal is serious stuff. We also have found that carrots are more effective than sticks, although we must occasionally resort to "sharp corrections," nearly always verbal. But if our interactions were limited to threats and punishments, then, in the words of a local cowboy, "That'd be a helluva way to have a horse."

Index

For the benefit of digital users, indexed terms that span two pages (e.g., 52–53) may, on occasion, appear on only one of those pages.

Bay of Pigs Invasion (1961) and,
134–35, 144
calls for US invasion of Cuba during,
145, 181
"Chicken" framing of, 181–82
DEFCON 2 status during, 145
De Gaulle's relationship with Kennedy
and, 26
nuclear deterrence theory and, 134–35,
144–45, 182, 196, 207
Soviet withdrawal of weapons from Cuba
and, 144–45, 181
submarine warfare and, 145–46
U2 planes and, 145, 146
US Congressional elections and, 185
US naval blockade of Cuba and, 144, 181
US nuclear weapons in Turkey and, 134–35,
144–45, 181
US pledge to not invade Cuba
following, 144–45
Cuban Revolution (1959), 136–37
Cuomo, Mario, 59–60
cuttlefish, 15–16
Czechoslovakia, 137

Dante Alighieri, 77–79
Dead Hand System (Soviet nuclear
arsenal), 182–83
death avoidance thinking, 81–88
death penalty. *See* capital punishment
deception. *See* dishonesty
De Gaulle, Charles, 26, 158
deimatic deeds, 15–16
The Denial of Death (Becker), 82–84
deterrence
assumptions of rationality and, 162–66
balance of power strategies and, 122
ballistic missile defense systems and, 128,
129, 188–90
brinkmanship and, 161–62
calls for the abandonment of nuclear
strategy of, 207–13
capital punishment and, 3, 5, 49, 51, 52,
54–63, 131
"Chicken" (game) and, 181–87, 194, 200
credibility and, 154–56, 161, 185–86
Cuban Missile Crisis (1962) and, 134–35,
144–45, 182, 196, 207

extended deterrence doctrine and,
158, 197–98
game theory and, 179
Great Wall of China and, 121
just war doctrine and, 174
Maginot Line and, 121
mutually assured destruction (MAD)
theory and, 128, 131, 155, 167, 174
nuclear deterrence and, 126–213
preemptive wars and, 197
punishment and, 5–6, 49–50, 122
Roman Empire's foreign policy
and, 122–23
DiAngelo, Robin, 108–9
Dick, Philip K., 179
Dickens, Charles, 108
"Dirge Without Music" (Millay), 88
dishonesty
deimatic deeds and, 15–16
empty threats and, 5, 17, 23, 25–26,
29, 154–55
evolutionary processes and, 16–17, 18
exaggeration and, 11
mimicry and, 15–16
District of Columbia v. Heller, 99
dominance
birds' coloration and, 30–31
escalation dominance and, 131, 160
interruption and, 41
vocal threats and, 34–37
Dominican Republic, 140
Donne, John, 87–88
Donohue III, John J., 58, 61–62
doves. *See* hawk-dove model of interaction
Drell, Sidney, 143
Dresden firebombing (1945), 173
Dr. Strangelove (film), 125, 146, 182–83,
205
Dukakis, Michael, 61
Dulles, John Foster, 161–62, 198–99
dung beetles, 23
Durkheim, Emile, 66
Duterte, Rodrigo, 101
dwarf mongooses, 25
The Dying Animal (Roth), 83

eastern collared lizards, 17
eavesdropping, 38–41

National Rifle Association (NRA), 90, 93, 94, 98, 100–1
Natural History (Pliny the Elder), 86
natural selection. *See under* evolution
Nero (emperor of Rome), 113
Nesse, Randy, 79
The Netherlands, 197–98
New York State, 55–56, 58–59, 61–62
New Zealand, 90–91, 95
Nicaragua, 123, 125, 136–37
Nice, Margaret, 36
Nicomachean Ethics (Aristotle), 74
Nitze, Paul, 208–9
Nixon, Richard, 127, 142, 163, 183, 199
Nobel, Alfred, 136
Norstad, Lauris, 198–99
North American Aerospace Defense Command (NORAD), 201–2
North Atlantic Treaty Organization (NATO)
    Able Archer 83 exercise (1983) and, 146–47, 149–51, 196
    expansion into Eastern Europe by, 103, 124
    limited war-fighting options and, 158
    US nuclear weapons in Europe and, 147, 148, 197–98
North Korea
    deterrence doctrine practiced by, 191–95
    Korean War (1950-53) and, 137–38, 192–93
    military warning system in, 152
    nuclear weapons and, 124, 129, 139–40, 142, 177, 192, 193–94
    Trump and, 124, 193, 194–95
    US antiballistic missile system plans and, 189–90
    USS Pueblo incident (1968) and, 138
Norwegian Rocket Incident (1995), 152, 201
nuclear weapons
    Able Archer 83 exercise (1983) and, 146–47, 149–51, 196
    accidents and near accidents with, 203–4, 206–7
    arms races in, 125, 142, 175–78, 190
    assumptions of rationality regarding use of, 162–66
    ballistic missile defense systems and, 128, 129, 188–90

business considerations regarding, 171–72
"Chicken" (game) and, 181–83, 186–87, 194, 200
coercive diplomacy and, 139–42
counterterrorism and, 142–43
Cuban Missile Crisis (1962) and, 130–33, 134–35, 137–38, 141, 158, 167, 169
Dead Hand System (Soviet Union) and, 182–83
delegation of military authority and potential of unauthorized use of, 204–6
deterrence doctrine and, 126–213
escalation dominance and, 160
fallout shelters and, 184
false alarms and, 148–52, 182, 201–3, 211
financial costs associated with building and maintaining, 177
Hiroshima and Nagasaki atomic attacks (1945) and, 128, 139, 142, 156–57, 173
just war doctrine and, 174
Korean War (1950-53) and, 137–38
limited nuclear war doctrine and, 156, 198–99
moral considerations regarding, 173–75
mutually assured destruction (MAD) theory and, 128, 131, 155, 167, 174
myths regarding, 130–36
neutron bombs and, 159
Nuclear Deterrence Summit (2019) and, 171–72
Operation RYaN (1981) and, 147–48
Pershing II missiles and, 147, 148
Prisoner's Dilemma and, 180
Reykjavik summit (1986) and, 189
scope of damage unleashed by, 130–31, 155, 156–58, 159
Trump presidency and, 163–65, 177–78, 183–84, 190–91, 194
US strategic triad and, 176–78, 205–6
Vietnam War (1954-75) and, 138–39, 141, 183, 198–99

Obama, Barack, 65, 98–99, 108, 123–24, 177–78, 187
Occupational Safety and Health Administration (OSHA), 115
The Odyssey (Homer), 73
Ogarkov, Nikolai, 150

Able Archer 83 exercise (1983) and, 146–47,
149–51, 196
Afghanistan War (1979-89) and,
136–37, 138–39
Cold War arms race and, 125, 176–77
Cold War threat of nuclear war and,
130–33, 134–35, 137–38, 141, 158,
167, 169
Cuban Missile Crisis (1962) and, 26,
134–35, 144, 181–82, 185
Hungary crackdown (1956) and, 137, 185
Operation RYaN (1981) and, 147–48
SS-20 intermediate-range missiles
and, 147–48
Yom Kippur War (1973) and, 199
Spain, 108–9
sparrows, 30–31, 36, 154
spiders, 7, 8
SS-20 intermediate-range missiles (Soviet
Union), 147–48
Stalin, Joseph, 137, 139, 141
stare threats, 7–8
Stevens, Wallace, 166
Stevenson, Adlai, 161
Stoics, 85–86, 88
stomatopods, 32–34
Stone, Michael H., 98
stotting, 21–22, 23
Strategic Defense Initiative, 128, 188, 189
strawberry poison frogs, 13
submarines
Cuban Missile Crisis (1962) and, 145–46
Trident submarines with nuclear weapons
and, 147, 174, 178
US nuclear strategic triad and, 176–77,
178, 206
Suez Crisis (1956), 138
suicide
guns in the United States and, 88, 92, 97,
98, 115
religious beliefs and, 66
Trump era and, 115
Sweden, 104–5
Syria, 125, 136, 137

Taiwan, 198–99
Taoism, 73
Tavris, Carol, 71–72

Tegmark, Max, 4–5
Tennyson, Alfred Lord, 8–9
terrorism
anxiety regarding, 1–2, 137
in Europe during twenty-first century,
138, 143
far-right groups' engagement in, 105
"Global War Against Terrorism" (US
policy, 2002-8), 123
nuclear deterrence theory and, 142–43
September 11 terrorist attacks (2001) and,
70–71, 138, 143
terror management theory (TMT), 83–84
Tertullian, 76
Texas, 54, 58–59, 92–93, 115
Thanatos (death instinct), 82
Thich Nhat Hanh, 85
"Things Come Apart" (Yeats), 106
Thirteen Days (Robert Kennedy), 145
"Thoughts for the Times on War and Death"
(Freud), 82
threat displays, 17–18, 23, 33, 40
threat inflation, 122–24
Thucydides, 161
Tinbergen, Niko, 9
Tisdale, Sallie, 85
Tokyo firebombing (1944-45), 173
Trident D5 missiles, 172
Trident submarines, 147, 174, 178
Truman, Harry S., 173
Trump, Donald
Affordable Care Act curtailed by, 115
capital punishment and, 64
counterterrorism policy and, 123
crime rates cited by, 89
gun policy and, 68
immigration policy and, 104–5,
108–9, 111
Iran policy and, 123–24, 187
Israel-Palestinian conflict and, 3
lying and braggadocio by, 25–26
national populism and, 101–2, 104–5,
106–7, 108–11, 114–15, 165
North Korea and, 124, 193, 194–95
nuclear weapons and, 163–65, 177–78,
183–84, 190–91, 194
Occupational Safety and Health
Administration curtailed by, 115